PROCESS PIPING DESIGN HANDBOOK

Volume One
The Fundamentals of Piping Design

Drafting and Design Methods for Process Applications

PROCESS PIPING DESIGN HANDBOOK

Volume One: The Fundamentals of Piping Design
Volume Two: Advanced Piping Design

PROCESS PIPING DESIGN HANDBOOK

Volume One
The Fundamentals of Piping Design

Drafting and Design Methods for Process Applications

Peter Smith

Gulf Publishing Company
Houston, Texas

Process Piping Design Handbook
Volume One: The Fundamentals of Piping Design

Copyright © 2007 by Gulf Publishing Company, Houston, Texas. All rights reserved. No part of this publication may be reproduced or transmitted in any form without the prior written permission of the publisher.

HOUSTON, TX:
Gulf Publishing Company
2 Greenway Plaza, Suite 1020
Houston, TX 77046

AUSTIN, TX:
427 Sterzing St., Suite 104
Austin, TX 78704

10 9 8 7 6 5 4 3 2 1

Library of Congress Cataloguing-in-Publication Data

Smith, Peter
Process piping design handbook, volume one:: the fundamentals of piping design / Peter Smith.
  v. cm. — (Process piping design series; 1)
  Includes bibliographical references and index.
  ISBN-13: 978-1-933762-04-3 (v. 1: acid-free paper)
  ISBN-10: 1-933762-04-7 (v. 1: acid-free paper)
  1. Pipelines—Design and construction. 2. Piping—Design and construction. 3. Piping—Computer-aided design. 4. Petroleum refineries—Equipment and supplies. I. Title.
  TA660.P55S65 2007
  621.8'672—dc22
                              2006038256

Printed in the United States of America
Printed on acid-free paper. ∞
Text design and composition by TIPS Technical Publishing, Inc.

*This book is dedicated to my son Stewart,
who is a constant source of inspiration and motivation to me.
He is sadly missed by his family and friends,
but he is never far away.*

Stewart Smith, Musician
March 11, 1972, to October 21, 2005

Contents

List of Figures xiii
List of Tables xvii
Foreword xix
Preface xxi

1 Piping Codes, Standards, and Specifications 1

   1.1 Introduction 1
   1.2 Definitions 2
   1.3 Codes 3
       1.3.1 American Society of Mechanical Engineers Boiler Pressure Vessel Codes 3
       1.3.2 American Society of Mechanical Engineers B31, Codes for Pressure Piping 14
   1.4 Standards and Specifications 24
       1.4.1 American Society of Mechanical Engineers 26
       1.4.2 American Petroleum Institute 30
       1.4.3 American Society for Testing and Materials 32
       1.4.4 American Society for Nondestructive Testing 39
       1.4.5 American Society for Quality 39
       1.4.6 American Welding Society 40
       1.4.7 American Water Works Association 41
       1.4.8 Copper Development Association 42
       1.4.9 Compressed Gas Association 43
       1.4.10 Canadian Standards Association 43

1.4.11 Expansion Joint Manufacturers Association   43
1.4.12 Manufacturers Standardization Society of the Valve and Fittings Industry   43
1.4.13 National Association of Corrosion Engineers   45
1.4.14 National Fire Protection Association   45
1.4.15 Pipe Fabrication Institute   46
1.4.16 Society of Automotive Engineers   48

2 Piping Components 49

2.1 Introduction to Piping Components   50
2.2 Pipe   51
   2.2.1   Pipe Sizes   62
   2.2.2   Pipe Ends   67
2.3 Pipe Fittings   70
   2.3.1   Butt-Weld End Fittings   72
   2.3.2   Socket-Weld and Threaded-End Fittings   73
   2.3.3   Flanged Joints   74
2.4 Valves   80
   2.4.1   Valve Codes and Standards   82
   2.4.2   Classification of Operation Valves   87
   2.4.3   Valve Classification   92
   2.4.4   Valve Components   94
2.5 Bolts and Gaskets (Fasteners and Sealing Elements)   101
   2.5.1   The Process of Joint Integrity   102
   2.5.2   Flange Joint Components   103
   2.5.3   The Flanged Joint System   111

3 Metallic Materials for Piping Components 115

3.1 Properties of Piping Materials   117
   3.1.1   Chemical Properties of Metals   117
   3.1.2   Mechanical Properties of Metals   118
   3.1.3   Elongation and Reduction of Area   119
   3.1.4   Physical Properties of Metals   120
3.2 Metallic Materials   122
3.3 Alloying of Steel   122

3.4 Types of Steel   125
   3.4.1   Mild (Low-Carbon) Steel   125
   3.4.2   Medium-Carbon Steel   125
   3.4.3   High-Carbon Steel   125
   3.4.4   High-Tensile Steel   125
   3.4.5   Stainless Steel   126
3.5 Steel Heat-Treating Methods   131
   3.5.1   Annealing   131
   3.5.2   Normalizing   132
   3.5.3   Hardening   132
   3.5.4   Tempering   132
3.6 Nonferrous Metals in Alloying   132
3.7 Material Specifications   133
   3.7.1   American Society for Testing and Materials   133
   3.7.2   Unified Numbering System of Ferrous Metals and Alloys   135

4   Roles and Responsibilities 137

4.1 The Lead Piping Engineer   138
4.2 Piping Materials Engineering Group   139
   4.2.1   Project Lead Piping Materials Engineer   140
   4.2.2   Senior Piping Materials Engineer   141
4.3 Piping Design Group   142
   4.3.1   Project Piping Area/Unit Supervisor (Squad Boss)   143
   4.3.2   Project Piping CAD Coordinator   143
   4.3.3   Project Piping Designers-Checkers   144
4.4 Piping Materials Control Group   145
   4.4.1   Project Lead Piping Materials Controller   146
   4.4.2   Project Piping Materials Controller   147
4.5 Piping Stress Engineering Group   148
   4.5.1   Project Lead Piping Stress Engineer   148
   4.5.2   Project Piping Stress Engineer   148
4.6 Other Engineering Disciplines Involved   150
   4.6.1   Process Engineering   151
   4.6.2   Mechanical Engineering   154

        4.6.3 Instrument Engineering 155
        4.6.4 Civil Engineering 155
        4.6.5 Structural Engineering 155

5 Projects 157
    5.1 Project Types 157
    5.2 Project Phases 160
        5.2.1 Feasibility Phase 160
        5.2.2 Conception Phase 160
        5.2.3 Front-End Engineering Development Phase 161
        5.2.4 Detailed Engineering Phase 162
        5.2.5 Construction Phase 166
        5.2.6 Precommissioning and Commissioning Phase 169
        5.2.7 Startup and Handover to the Owner 169

6 Fabrication, Assembly, and Erection..................... 171
    6.1 Codes and Standards Considerations 172
    6.2 Fabrication Materials for Piping Systems 172
    6.3 Fabrication Drawings 173
    6.4 Fabrication Activities 174
        6.4.1 Cutting 174
        6.4.2 Beveling 174
        6.4.3 Forming 175
        6.4.4 Bending 175
    6.5 Welding 178
        6.5.1 Welding Processes 179
        6.5.2 Preheating 181
        6.5.3 Heat Treatment 181
    6.6 Brazing and Soldering 183
    6.7 Protection of Carbon Steel in Corrosive Services 183
        6.7.1 Corrosion Allowance 184
        6.7.2 Internal Galvanizing of Pipe and Piping Systems 184
        6.7.3 Pipe Cladding 186

6.8 Assembly and Erection 186
    6.8.1 Alignment 186
    6.8.2 Flanged Joints 188
    6.8.3 Threaded Joints 188

7 Inspection and Testing 191

7.1 Piping Codes 191
7.2 Types of Examination 196
    7.2.1 Visual Examination 197
    7.2.2 Liquid Penetrant Examination 197
    7.2.3 Magnetic Particle Examination 198
    7.2.4 Radiographic Examination 199
    7.2.5 Ultrasonic Examination 200
7.3 Testing of Piping Systems 200
    7.3.1 Some Limitations on the Pressure Testing of Piping Systems 201
    7.3.2 Special Provisions for Testing 202
    7.3.3 Preparation for Leak Test 202
7.4 Leak-Testing Methods 202
    7.4.1 Hydrostatic Leak Test 202
    7.4.2 Pneumatic Leak Test 204
    7.4.3 Combination Hydrostatic-Pneumatic Leak Test 204
    7.4.4 Initial Service Leak Testing 205
7.5 Choice of Testing Medium 205
7.6 Test Pack 206
7.7 Punch List 206

A Listed Material 209

    A.0.1 Ferrous Metals 209
    A.0.2 Nonferrous Material 213
A.1 American Petroleum Institute 217

B General Engineering Data 219

Index 231

List of Figures

Figure P–1	An example of a snap shot taken from a 3D CAD model. (Printed with the permission of Bentley Systems Incorporated.)	xxii
Figure P–2	A 3D model of five gas compressors showing inlet and outlet piping. (Printed with the permission of Bentley Systems Incorporated.)	xxii
Figure P–3	A 3D model of one gas compressor showing the pipe work in more detail. (Printed with the permission of Bentley Systems Incorporated.)	xxiii
Figure 1–1	Onshore pipelines are designed to ASME B31.4 for liquid transportation and ASME B31.8 for gas transmission. (Printed with the permission of Bentley Systems Incorporated.)	17
Figure 2–1	A 3D model that shows a variety of components used to pipe up a vertical vessel. (Printed with the permission of Bentley Systems Incorporated.)	51
Figure 2–2	Bevels for wall thickness over 3 mm (0.12 in.) to 22 mm (0.88 in.), inclusive. (Printed with the permission of ASME.)	71
Figure 2–3	Well bevel details for wall thickness over 22 mm (0.88 in.). (Printed with the permission of ASME.)	71

Figure 2–4	Recommended gap for socket weld fit up, prior to welding.. 77
Figure 2–5	Gate valve, bolted bonnet, outside screw and yoke. (Printed with the kind permission of OMB Valves, spa, Italy.).. 88
Figure 2–6	Globe valve, bolted bonnet, outside screw and yoke. (Printed with the kind permission of OMB Valves, spa, Italy.).. 89
Figure 2–7	Piston, ball and swing type threaded or socket weld ends. (Printed with the kind permission of OMB Valves, spa, Italy.).. 90
Figure 2–8	Piston, ball and swing type flanged ends. (Printed with the kind permission of OMB Valves, spa, Italy.) 91
Figure 2–9	Tilting-disc check valve, pressure seal bonnet. (Printed with the kind permission of Valvosider, srl, Italy.)... 92
Figure 2–10	This is an exploded view of a wafer-type dual-plate check valve that will be held between two flanges: double flanged (top) and solid lug type (bottom). (Printed with the kind permission of Goodwin International, Ltd., United Kingdom.) 93
Figure 2–11	Pinch valve with the components of construction. (Printed with the permission of Resistoflex.) 94
Figure 2–12	Ball valve, flanged, side entry ASME class 150 full bore and reduced bore. (Printed with the kind permission of OMB Valves, spa, Italy.) 95
Figure 2–13	Butterfly valve: (top) 84" class 150 butterfly valve for water service and (bottom) a general arrangement for a butterfly valve. (Printed with the permission of Curtiss Wright Controls.).. 96
Figure 2–14	Part section through a flanged plug valve. (Printed with the permission of Durco.)................................. 97

Figure 2–15	Section through a rubber-lined flanged diaphragm valve. (Printed with the permission of Saunders.)	98
Figure 4–1	An example of an isometric extracted from a 3D CAD model. (Printed with the permission of Bentley Systems Incorporated.)	146
Figure 4–2	An example of a P&ID legend sheet. (Printed with the permission of Bentley Systems Incorporated.)	153
Figure 4–3	An example of a P&ID. (Printed with the permission of Bentley Systems Incorporated.)	154
Figure 5–1	A refinery designed to ASME B31.3. (Printed by permission of Bentley Systems Incorporated.)	159
Figure 5–2	A 3D CAD model of a vessel, showing foundation, inlet, outlet process piping, utility pipe work, and platforms. (Printed with the permission of Bentley Systems Incorporated.)	164
Figure B–1	API Standard 600 trim materials. (Printed with the kind permission of Valvosider, srl, Italy.)	220
Figure B–2	Conversion chart—pressure and temperature. (Printed with the kind permission of OMB Valves, spa, Italy.)	221
Figure B–3	Material equivalents. (Printed with the kind permission of OMB Valves, spa. Italy.)	222
Figure B–4	Pressure and temperature ratings—material groups 1.1 and 1.9. (Printed with the kind permission of OMB Valves, spa, Italy.)	223
Figure B–5	Pressure and temperature ratings—material groups 1.10 and 2.2. (Printed with kind permission of OMB Valves, spa, Italy.)	224
Figure B–6	Conversion chart dimensions—U.S. customary units to metric units. (Printed with kind permission of OMB Valves, spa, Italy.)	225

Figure B–7	Standard components of construction for forged small-bore gate, globe, and check valves. (Printed with the kind permission of OMB Valves, spa, Italy.)	226
Figure B–8	Conversion chart dimensions—U.S. customary units to metric units. (Printed with kind permission of OMB Valves, spa, Italy.)	227
Figure B–9	Standard components of construction for forged small-bore gate, globe, and check valves. (Printed with the kind permission of OMB Valves, spa, Italy.)	228
Figure B–10	Standard materials of construction for small-bore gate, globe, and check valves. (Printed with the kind permission of OMB Valves, spa, Italy.)	229
Figure B–11	Threaded end connections to ASME B1.20.1. (Printed with the kind permission of OMB Valves, spa, Italy.)	230

List of Tables

Table 2–1	Basic Quality Factors for Longitudinal Weld Joints in Pipes, Tubes, and Fittings E_j	53
Table 2–2	*Diamètre Nominal* (DN) and Nominal Pipe Size (NPS)	64
Table 2–3	Types of Operation Valves	87
Table 3–1	Austenitic Steel Grades	127
Table 3–2	Austenitic Steel Grades	128
Table 3–3	Ferritic Steel Grades	129
Table 3–4	Nonferrous Metals Used in Alloying Steel	133
Table 3–5	UNS Series	136
Table 6–1	ASTM Standards Relating to Hot-Dip Galvanizing and Hot-Dip Galvanized Materials	187

Foreword

I first became involved with process piping design over 30 years ago, while still an engineering student. During summer breaks I worked for an engineering company, drafting piping isometrics for power plant projects. The tools back then were a T-square, triangle, and a pencil. With adequate drafting skills and a few days to learn the arcane symbology on piping drawings, I was productive enough to earn my keep. Proud as I was of my new piping skills, I quickly learned just how much I did not know. Even after graduating, armed with my engineering degree, I realized I still had much to learn. Applying sound design practices to the multitude of process systems; understanding the many codes and standards to which a piping system must comply; selecting the right component and the right material for a specific application; creating piping systems that can be effectively supported, fabricated, constructed, operated and maintained; and on and on—all these were in my future.

In the intervening years, I have been involved in the development and application of information technology to the design of piping systems. Computer-aided drafting, 3D computer modeling, design visualization, engineering analysis, and construction simulation, now fairly common in the design of piping systems, have gone a long way to improve the productivity, quality, and schedules associated with piping system design and engineering. But, in spite of the power of these tools, a tremendous amount of knowledge still is required to be effective in designing process piping systems, just as was the case 30 years ago, when I started out working with a pencil.

Providing this critical, fundamental piping knowledge is Peter Smith's mission here in the *Process Piping Design Manual*. His focus is on *practice* not *technological tools*. In these two volumes, he covers a broad spectrum of knowledge and information, all of which is critical to sound process piping design: codes and standards, piping components, design practices and processes, mechanical equipment, piping materials, and more. The *Process Piping Design Manual* is an excellent learning resource for those new to process piping design. For the more experienced, it is a reliable reference to which one can return again and again.

These two manuals will serve as a platform for piping engineers who wish to progress into their chosen sectors, whether it be piping layouts, materials, or stress.

The tools we apply to process piping design will continue to evolve and improve, becoming more productive and cost effective along the way. But, regardless of the quality of our tools, it is the *practice* of process piping design that determines how successfully we apply these tools. Peter Smith provides a real service to the piping industry through the sound and practical knowledge he imparts in these manuals.

—A. B. Cleveland, Jr.
Senior Vice President
Bentley Systems Incorporated

Preface

In the CAD era, the rapid input and extraction of technical information far exceeds the manual drafting process previously used to design process plants. It is very important that the individuals responsible for CAD modeling and manipulation of this information understand the basic principles of piping design.

However, the industry has not stood still. In one area, there has been a dramatic change: the design of process plants and the visual representation of piping systems. This change was driven by information technology (IT) and altered the very way process plants are conceived, visualized, developed, designed, constructed, and maintained. Computer-aided design (CAD) has now almost totally replaced manual draftsmanship (see Figure P–1).

This radical shift came about in the last 20 years, and it accelerated in the last 10 years to such an extent that the "old school" of manually designing a piping system in drawings, using pencil and paper, has been replaced by digital representation and the creation of 2D CAD drawings and the more advances 3D model (see Figures P–1, P–2, and P–3).

Engineers and designers who entered the industry post 1990 will be less aware of manual representation of piping systems, and certain individuals have never been asked to "draw" a general arrangement or an isometric. The age of digital representation of technical drawings is a revelation, and it does increase the speed of the design phase of a project if all of the information is available. It also increases the speed at which the individual can make mistakes, and there are no limits to the severity of these mistakes.

Figure P-1 *An example of a snap shot taken from a 3D CAD model. (Printed with the permission of Bentley Systems Incorporated.)*

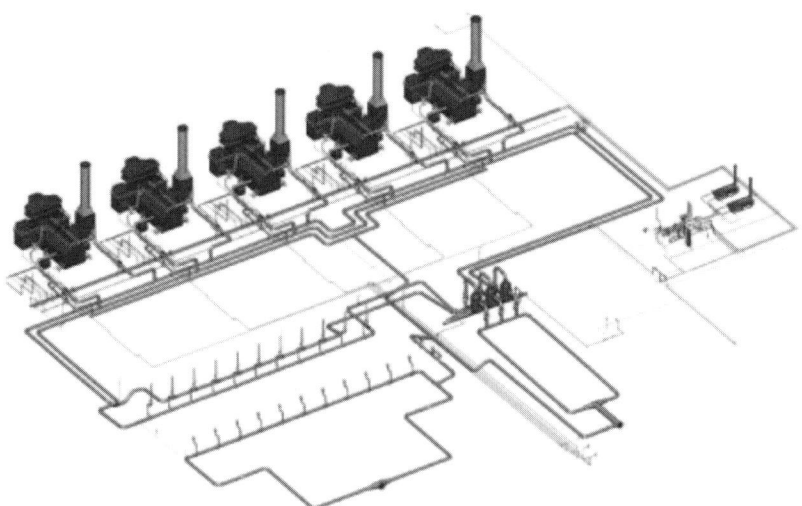

Figure P-2 *A 3D model of five gas compressors showing inlet and outlet piping. (Printed with the permission of Bentley Systems Incorporated.)*

Figure P–3 *A 3D model of one gas compressor showing the pipe work in more detail. (Printed with the permission of Bentley Systems Incorporated.)*

At this time of writing, "intelligent" software is being discussed in the industry, however this facility is useful only if it is applied by an "intelligent" operator. In many cases, the IT skills of CAD operators far exceed their piping skills, and the objective is to get a healthy balance, where an engineer or designer not only understands how to model quickly, but why he or she is modeling a specific component or layout.

Speed and accuracy should be the goal.

When I first came in contact with computers in the mid 1970s, there was a saying, "garbage in—garbage out." This still applies, and it will do so for the foreseeable future.

The objective of *The Design of Process Piping Systems*, volume one, is to introduce the reader to the fundamental rules of the subject. Most of these rules originate from the various industry codes, such as the American Society of Mechanical Engineers (ASME) B31 series; and in many cases, these rules are mandatory requirements, which when applied result in a plant that can be constructed and operated safely for its projected lifetime.

Other rules or standards are required to allow the plant to be designed and constructed uniformly with fittings with predetermined dimensions and manufactured of materials with known chemical composition and predictable mechanical strength. It is essential that an engineer or designer understands the basic requirements and the background to the options that are available during the design phase.

I reference process piping systems designed to one of the ASME B31 series of codes, with an emphasis on ASME B31.3—Power Piping. Piping is a very large subject and it would take numerous volumes to cover it in its entirety. Also, many specialist sectors require in-depth reading, such as metallurgy, welding, corrosion, inspection, and valves.

What I attempted to do is to give the reader fundamental information on the subject of piping that can be expanded on by further reading and "on the job" experience.

Volume one covers the following topics:

- Piping codes, standards, and specifications.
- Piping components.
- Metallic materials for piping components.
- Roles and responsibilities.
- Projects.
- Fabrication, assembly, and erection.
- Inspection and testing.

In volume two, I cover the various commonly used piping systems and the hazardous and nonhazardous fluids that they transport. I also discuss the numerous types of piping layouts required for the different types of process equipment found in a plant.

CHAPTER 1

Piping Codes, Standards, and Specifications

In the new computer-aided design (CAD) era, the compliance to industry codes, standards, and specifications remains essential for the successful completion of a process facility, safe operation, and the satisfaction of health, safety, and environmental (HSE) requirements. The chapter is divided into the following sections:

1.1 Introduction

1.2 Definitions

1.3 Codes

1.4 Standards and Specifications

1.1 Introduction

Compliance to a code generally is mandatory, imposed by regulatory and enforcement agencies or their representatives. Also, the insurance company for the facility requires the owner to comply with the requirements of the relevant code or codes to ensure the safety of the workers and the general public. Compliance to standards normally is required by the rules of the applicable code or the purchaser's specification.

A vast majority of these codes, standards, and specifications have their origins in the United States, because initially this is where the most oil and gas activity was based. This is not likely to change in the near future; however, in recent years, there has been an increase in the alignment with ISO, and this is likely to increase.

Despite the strength of U.S. codes, standards, and specifications, similar documents from other engineering centers should not be ignored, like British standards (UK), DIN (Germany), AFNOR (France), JIS (Japan), and others.

1.2 Definitions

A *code* identifies the general requirements for the design, materials, fabrication, erection, test, and inspection of process piping systems. For example, ASME B31.3—Process Piping is classified as a design code. This is the most commonly used international design code for process plants.

A *standard* contains more-detailed design and construction parameters and standard dimensional and tolerance requirements for individual piping components, such as various types of valves, pipe, tee, flanges, and other in-line items to complete a piping system. For example, ASME B16.5, Pipe Flanges and Flanged Fittings, is classified as a dimensional standard, but it also references ASTM material specifications.

A *specification*, as the word implies, gives more specific information and data on the component; and ASTM's are considered to be material specifications, although they sometimes are ambiguously called *standard specifications*. ASTM A105 is the "standard specification for carbon steel forgings for piping applications."

To conclude and combine these definitions, ASME B31.3 is a design *code*, with flanges designed to the ASME B16.5 *standard*, which are constructed to the material *specification* ASTM A105.

It is not uncommon for even experienced personnel to get the definitions of these three types of document mixed up, and it is important to comprehend the distinct differences.

1.3 Codes

A regulatory organization imposes mandatory compliance to a code, from the basic design through to mechanical completion and eventual hand-over of a plant to the operator. For example, ASME B31. 3, Process Piping, is the refinery code. The insurer of the plant will make this a contractual requirement to ensure safety for personnel and plant during construction, commissioning, and ongoing operation.

The codes, standards, and specifications that relate to piping systems and piping components are published by various organizations. These organizations have committees comprising representatives from industry associations, manufacturers, EPC contractors, end users/operators, government bodies, insurance companies, and other interested groups.

A committee is responsible for maintaining, updating, and revising the codes, standards, and specifications, taking into consideration all technological developments, research, experience feedback from end users, and any changes in referenced codes, standards, specifications, or regulations.

The oil and gas industry has been established for many years, and changes to industry codes are generally negligible. Periodically, revisions are published, listing amendments that have been made to the document. It is essential that engineers and designers who work regularly with the document use the latest edition.

With regard to referencing a particular edition, issue, addendum, or revision of a code or standard, the piping engineer must be aware of the national, state, provincial, and local laws and regulations governing its interpretation in addition to the commitments made by the owner and the limitations delineated in the code or standard.

1.3.1 American Society of Mechanical Engineers Boiler Pressure Vessel Codes

The boiler pressure vessel (BPV) section covers major codes and standards related to piping. Some of these codes and standards are discussed briefly, whereas others are listed for convenience of reference.

American Society of Mechanical Engineers

The American Society of Mechanical Engineers (ASME) is one of the leading engineering organizations in the world. It develops and publishes engineering codes and standards. The ASME established a committee in 1911 to formulate rules for the construction of steam boilers and other pressure vessels. This committee, now known as the ASME Boiler and Pressure Vessel Committee, is responsible for the ASME boiler and pressure vessel code. In addition, the ASME has established committees that develop many other codes and standards, such as the ASME B31 code for pressure piping.

ASME Boiler and Pressure Vessel Code

The ASME Boiler and Pressure Vessel Code comprises 12 sections:

> Section I, Power Boilers.
>
> Section II, Material Specifications.
>
> Section III, Rules for Construction of Nuclear Power Plant Components.
>
> - Division 1, Nuclear Power Plant Components.
>
> - Division 2, Concrete Reactor Vessel and Containments.
>
> - Division 3, Containment Systems and Transport Packaging for Spent Nuclear Fuel and High-Level Radioactive Waste.
>
> Section IV, Heating Boilers.
>
> Section V, Nondestructive Examination.
>
> Section VI, Recommended Rules for Care and Operation of Heating Boilers.
>
> Section VII, Recommended Rules for Care of Power Boilers.
>
> Section VIII, Pressure Vessels.
>
> - Division 1, Pressure Vessels.
>
> - Division 2, Pressure Vessels (Alternative Rules).

- Division 3, Alternative Rules for Construction of High-Pressure Vessels.

Section IX, Welding and Brazing Qualifications.

Section X, Fiber-Reinforced Plastic Pressure Vessels.

Section XI, Rules for In-Service Inspection of Nuclear Power Plant Components.

Section XII, Rules for Construction and Continued Service of Transport Tanks.

Code Cases: Boilers and Pressure Vessels.

Code Cases: Nuclear Components.

ASME Section I, Power Boilers

This ASME section provides requirements for all methods of construction of power, electric, and miniature boilers; high-temperature water boilers used in stationary service; and power boilers used in locomotive, portable, and traction service. Rules pertaining to use of the V, A, M, PP, S, and E code symbol stamps are included. The rules are applicable to boilers in which steam or other vapor is generated at pressures exceeding 15 psig and high-temperature water boilers intended for operation at pressures exceeding 160 psig or temperatures exceeding 250°F. Superheaters, economizers, and other pressure parts connected directly to the boiler without intervening valves are considered part of the scope of Section I.

ASME Section II, Material Specifications (Scope)

ASME Section II consists of four parts, three of which contain material specifications and the fourth the properties of materials listed previously.

Part A, Ferrous Material Specifications.

Part B, Nonferrous Material Specifications.

Part C, Specifications for Welding Rods, Electrodes, and Filler Metals.

Part D, Properties

Practical Guide to ASME Section II.

Part A, Ferrous Material Specifications, provides material specifications for ferrous materials adequate for safety in the field of pressure equipment. These specifications contain requirements and mechanical properties, test specimens, and methods of testing. They are designated by SA numbers and are derived from ASTM A specifications.

Part B, Nonferrous Material Specifications, provides material specifications for nonferrous materials adequate for safety in the field of pressure equipment. These specifications contain requirements for heat treatment, manufacture, chemical composition, heat and product analyses, mechanical test requirements, and mechanical properties, test specimens, and methods of testing. They are designated by SB numbers and derived from ASTM B specifications.

Part C, Specifications for Welding Rods, Electrodes, and Filler Metals, provides material specifications for the manufacture, acceptability, chemical composition, mechanical usability, surfacing, testing requirements and procedures, operating characteristics, and intended uses for welding rods, electrodes, and filler metals. These specifications are designated by SFA numbers and derived from AWS specifications.

Part D, Properties, provides tables of design stress values, tensile and yield strength values, and tables and charts of material properties. Part D facilitates ready identification of specific materials to specific sections of the boiler and pressure vessel code. Part D's appendices contain criteria for establishing allowable stress, the bases for establishing external pressure charts, and information required for approval of new materials.

Subpart 1 contains allowable stress and design stress intensity tables for ferrous and nonferrous materials for pipe, fittings, plates, bolts, and so forth. In addition, it provides tensile strength and yield strength values for ferrous and nonferrous materials and lists factors for limiting permanent strain in nickel, high-nickel alloys, and high-alloy steels.

Subpart 2 of Part D has tables and charts providing physical properties, such as the coefficient of thermal expansion, moduli of elasticity, and other technical data needed for the design and construction of pressure-containing components and their supports made from ferrous and nonferrous materials.

Section III, Rules for Construction of Nuclear Facility Components

Subsection NCA contains general requirements for divisions 1 through 3:

    Division 1.

        Subsection NB, Class 1 Components.

        Subsection NC, Class 2 Components.

        Subsection ND, Class 3 Components.

        Subsection NE, Class MC Components.

        Subsection NF, Supports.

        Subsection NG, Core Support Structures.

        Subsection NH, Elevated Temperature Service.

        Appendices.

    Division 2.

        Code for Concrete Reactor Vessels and Containments.

    Division 3.

        Containment Systems and Transportation Packaging for Spent Nuclear Fuel and High Level Radioactive Waste.

Division 1, Subsection NCA, General Requirements Divisions 1 and 2, provides tables of design stress values, tensile and yield strength values, and tables and charts of material properties. Part D facilitates ready identification of specific materials to specific sections of the boiler and pressure vessel code. Part D's appendices contain criteria for establishing allowable stress, the bases for establishing external pressure charts, and information required for approval of new materials.

Subsection NB, Class 1 Components, contains requirements for the material, design, fabrication, examination, testing, and overpressure protection of items intended to conform to the requirements for Class 1 construction. The rules of subsection NB cover the requirements for assuring the structural integrity of items.

Subsection NC, Class 2 Components, contains requirements for the material, design, fabrication, examination, testing, and overpressure protection of items intended to conform to the requirements for Class 2 construction. The rules of subsection NC cover the requirements for assuring the structural integrity of items.

Subsection ND, Class 3 Components, contains requirements for the material, design, fabrication, examination, testing, and overpressure protection of items intended to conform to the requirements for Class 3 construction. The rules of subsection ND cover the requirements for assuring the structural integrity of items.

Subsection NE, Class MC Components, contains requirements for the material, design, fabrication, examination, inspection, testing, and overpressure protection of metal containment vessels intended to conform to the requirements for class MC construction. The rules of subsection NE cover the requirements for assuring the structural integrity of the metal containment vessel.

Subsection NF, Supports, contains requirements for the material, design, fabrication, and examination of supports intended to conform to the requirements for classes 1, 2, 3 and MC construction. Nuclear power plant supports for which rules are specified in this subsection are those metal supports designed to transmit loads from the pressure retaining barrier of the component of piping to the load-carrying building structure. In some cases, intervening elements in the component support load path may not be constructed to the rules of this section, such as diesel engines, electric motors, valve operators, coolers, and access structures.

Subsection NG, Core Support Structures, contains requirements for the material, design, fabrication, and examination required in the manufacture and installation of core support structures. Core support structures are those structures or parts of structures designed to provide direct support or restraint of the core (fuel and blanket assemblies) within the reactor pressure vessel.

Subsection NH, Class 1 Components with Elevated Temperature Service, contains requirements for materials, design, fabrication, examination, testing, and overpressure relief of class 1 components, parts, and appurtenances expected to function even when metal temperatures exceed those covered by the rules and stress limits of subsection NB and Tables 2A, 2B, and 4 of Section II, Part D, Subpart 1.

The appendices for Section III, Division 1 (Subsection NCA through NG) and Division 2 include a listing of design and design analysis methods and information and data report forms. The appendices are referenced by and an integral part of Subsections NCA through NG and Division 2.

Division 2, Code Concrete Reactor Vessels and Containment, contains requirements for the material, design, construction, fabrication, testing, examination, and overpressure protection of concrete reactor vessels and concrete containment structures, prestressed or reinforced. These requirements are applicable only to those components designed to provide a pressure retaining or containing barrier. They are not applicable to other support structures, except as they directly affect the components of the systems. This section contains appendices for division 2 construction.

Division 3, Containments for Transportation and Storage, contains the rules of division 3 that constitute requirements for the design and construction of the containment system of a nuclear spent fuel or high-level radioactive waste transport packaging.

ASME Section IV, Rules for Construction of Heating Boilers

This ASME BPV subsection provides requirements for the design, fabrication, installation, and inspection of steam-generating boilers and hot-water boilers intended for low-pressure service that are directly fired by oil, gas, electricity, or coal. It contains appendices that cover approval of new material, methods of checking the safety valve and safety relief valve capacity, examples of methods of checking safety valve and safety relief valve capacity, examples of methods of calculation and computation, definitions relating to boiler design and welding, and quality control systems. Rules pertaining to use of the H, HV, and HLW code symbol stamps are also included.

BPVC-V, 2004, BPVC Section V, Nondestructive Examination (Scope)

This ASME BPV section contains requirements and methods for nondestructive examination that are referenced and required by other code sections. It also includes manufacturer's examination responsibilities, duties of authorized inspectors, and requirements for qualification of personnel, inspection, and examination. Examination methods are intended to detect surface and internal discontinuities in materials, welds, and fabricated parts, and components. A glossary of related terms is included.

BPVC-VI, 2004, BPVC Section V, Recommended Rules for the Care and Operation of Heating Boilers (Scope)

This ASME BPV section covers general descriptions, terminology, and operation guidelines applicable to steel and cast-iron boilers limited to the operating ranges of Section IV, Heating Boilers. It includes guidelines for associated controls and automatic fuel-burning equipment. Illustrations show typical examples of available equipment. Also included is a glossary of terms commonly associated with boilers, controls, and fuel-burning equipment.

BPVC-VII, 2004, BPVC Section VII, Recommended Guidelines for the Care of Power Boilers

The purpose of these guidelines is to promote safety in the use of stationary, portable, and traction-type heating boilers. This section provides such guidelines to assist operators of power boilers in maintaining their plants as safely as possible. Emphasis has been placed on industrial type boilers because of their extensive use.

BPVC-VIII, 2004, BPVC Section VIII, Rules for Construction of Pressure Vessels

> Division 1, Rules for the Construction of Pressure Vessels.
>
> Division 2, Alternative Rules for the Construction of Pressure Vessels.
>
> Division 3, Alternative Rules for the Construction of High Pressure Vessels.

Division 1, Rules for Construction of Pressure Vessels, of ASME BPV Section VIII provides requirements applicable to the design, fabrication, inspection, testing, and certification of pressure vessels operating at either internal or external pressures exceeding 15 psig. Such pressure vessels may be fired or unfired. Specific requirements apply to several classes of material used in pressure vessel construction and fabrication methods, such as welding, forging, and brazing. It contains appendices detailing supplementary design criteria, nondestructive examination, and inspection acceptance standards. Rules pertaining to the use of the U, UM, and UV code symbol stamps are included.

Division 2, Alternative Rules for the Construction of Pressure Vessels, provides requirements applicable to the design, fabrication, inspection, testing, and certification of pressure vessels operating at either internal or external pressures exceeding 15 psig. Such vessels may be fired or unfired. This pressure may be obtained from an external source, by the application of heat from a direct or indirect source, or any combination thereof. These rules provide an alternative to the minimum requirements for pressure vessels under division 1 rules. In comparison to division 1, division 2 requirements on materials, design, and nondestructive examination are more rigorous; however, higher design stress intensity values are permitted. Division 2 rules cover only vessels to be installed in a fixed location for a specific service, where operation and maintenance control is retained during the useful life of the vessel by the user who prepares or causes to be prepared the design specifications. These rules may also apply to human occupancy pressure vessels, typically in the diving industry. Rules pertaining to the use of the U2 and UV code symbol stamps are included.

Division 3, Alternative Rules for the Construction of High Pressure Vessels, provides requirements applicable to the design, fabrication, inspection, testing, and certification of pressure vessels operating at either internal or external pressures generally above 10,000 psig. Such vessels may be fired or unfired. This pressure may be obtained from an external source, a process reaction, by the application of heat from a direct or indirect source or any combination thereof. Division 3 rules cover vessels intended for a specific service and installed in a fixed location or relocated from work site to work site between pressurizations. The operation and maintenance control is retained during the useful life of the vessel by the user who prepares or causes to be prepared the design specifications. Division 3 does not establish maximum pressure limits for either section VIII or divisions 1 or 2 nor minimum pressure limits for this division. Rules pertaining to the use of the UV3 code symbol stamps are included.

BPVC-IX, 2004, BPVC Section IX, Welding and Brazing Qualifications (Scope)

This ASME BPV section contains rules relating to the qualification of welding and brazing procedures as required by other code sections for component manufacture. It also covers rules relating to the qualification and requalification of welders, brazers, and welding and brazing operators in order that they may perform welding or brazing as required by other code sections in the manufacture of components. Welding and brazing data cover essential and nonessential variables specific to the welding or brazing process used.

BPVC-X, 2004, BPVC Section X, Fiber-Reinforced Plastic Pressure Vessels (Scope)

This ASME BPV section provides requirements for construction of a fiber-reinforced plastic (FRP) pressure vessel in conformance with a manufacturer's design report. It includes production, processing, fabrication, inspection, and testing methods required for the vessel. Section X includes two classes of vessel design: class 1, a qualification through the destructive test of a prototype, and class 2, mandatory design rules and acceptance testing by nondestructive methods. These vessels are not permitted to store, handle, or process lethal fluids. Vessel fabrication is limited to the following processes: bag molding, centrifugal casting and filament winding, and contact molding. General specifications for the glass and resin materials and minimum physical properties for the composite materials are given.

BPVC-XI, 2004, BPVC Section XI, Rules for In-Service Inspection of Nuclear Power Plant Components (Scope)

This ASME BPV section contains divisions 1 and 3 in one volume and provides rules for the examination, in-service testing and inspection, and repair and replacement of components and systems in light water-cooled and liquid-metal-cooled nuclear power plants. The division 2 rules for inspection and testing of components of gas-cooled nuclear power plants were deleted in the 1995 edition. With the decommissioning of the only gas-cooled reactor to which these rules apply, there is no apparent need to continue publication of division 2. Application of this section of the code begins when the requirements of the construction code have been satisfied. The rules of this section constitute requirements to maintain the nuclear power plant while in operation and to return the plant to service following plant outages and repair or replacement activities. The rules require a mandatory program of scheduled examinations, testing, and inspections to evidence adequate safety. The method of nondestructive examination to be used and flaw size characterization are contained within this section.

BPVC-XII, 2004, BPVC Section XII, Rules for Construction and Continued Service of Transport Tanks (Scope)

This ASME BPV section covers requirements for construction and continued service of pressure vessels for the transportation of dangerous goods via highway, rail, air, or water at pressures from full vacuum to 3000 psig and volumes greater than 120 gallons. *Construction* is an all-inclusive term comprising materials, design, fabrication, examination, inspection, testing, certification, and overpressure protection. *Continued service* is an all-inclusive term referring to inspection, testing, repair, alteration, and recertification of a transport tank that has been in service. This section contains modal appendices containing requirements for vessels used in specific transport modes and service applications. Rules pertaining to the use of the T code symbol stamp are included.

Code Cases, Boilers and Pressure Vessels (Scope)

This ASME BPV volume contains provisions adopted by the Boiler and Pressure Vessel Committee that cover all sections of the code other than section III, divisions 1 through 3, and section XI to provide, when the need is urgent, rules for materials or constructions not covered by existing code rules. Code case revisions in the form of supplements are sent automatically to purchasers up to the publication of the 2007 code.

Code Cases, Nuclear Components (Scope)

This ASME BPV volume contains provisions adopted by the Boiler and Pressure Vessel Committee that cover section III, divisions 1 through 3, and section XI to provide, when the need is urgent, rules for materials or constructions not covered by existing code rules. Code case revisions in the form of supplements are sent automatically to purchasers up to the publication of the 2007 code.

1.3.2 American Society of Mechanical Engineers B31, Codes for Pressure Piping

Before design engineering work can commence on a process unit, it must be established which international codes, standards, and specifications apply to the project. Without these essential documents in place, it is impossible to deliver a project that meets international safety levels and engineering quality necessary for the plant owner to be granted an operating license.

Project B31 was started in March 1926 and the first edition of *American Standard Code for Pressure Piping* was published in 1935. Over the years, several sections of the code for pressure piping were published to cover various sectors of the energy industry. Since December 1978, the American National Standards Committee B31 was reorganized as the ASME Code for Pressure Piping B31 Committee under procedures developed by the ASME and accredited by American National Standards Institute (ANSI).

Presently, the following sections of ASME B31, Code for Pressure Piping, are published:

> ASME B31.1, Power Piping.
>
> ASME B31.2, Fuel Gas Piping.
>
> ASME B31.3, Process Piping.
>
> ASME B31.4, Liquid Transportation Systems for Hydrocarbons, Liquid Petroleum Gas, Anhydrous Ammonia, and Alcohol.
>
> ASME B31.5, Refrigeration Piping.
>
> ASME B31.8, Gas Transmission and Distribution Piping Systems.
>
> ASME B31.8S, Managing System Integrity of Gas Pipelines.
>
> ASME B31.9, Building Services Piping.
>
> ASME B31.11, Slurry Transportation Piping Systems.
>
> B31G, Manual for Determining Remaining Strength of Corroded Pipelines.
>
> ASME B31, Standards of Pressure Piping.

The group of ASME B31 codes, previously known as ANSI B31, covers pressure piping, was created by the American Society of Mechanical Engineers, and includes power piping, fuel gas piping, process piping, pipeline transportation systems for liquid hydrocarbons and other liquids, refrigeration piping and heat transfer components, and building services piping.

B31.1, Power Piping

This ASME code prescribes minimum requirements for the design, materials, fabrication, erection, test, and inspection of power and auxiliary service piping systems for electric generation stations, industrial institutional plants, central and district heating plants.

The code covers boiler external piping for power boilers and high-temperature, high-pressure water boilers in which steam or vapor is generated at a pressure of more than 15 psig and high-temperature water generated at pressures exceeding 160 psig or temperatures exceeding 250°F.

B31.2, 1968 Fuel Gas Piping

This ASME code covers the design, fabrication, installation, and the testing of piping systems for fuel gases used in buildings and between buildings, from the outlet of the meter to the first pressure valve. Note: This code was withdrawn as an ANSI standard in 1968. A portion of the B31.2 code requirements have been incorporated into B31.8.

B31.3, Process Piping

This ASME code covers rules for piping typically found in petroleum refineries; chemical, pharmaceutical, textile, paper, semiconductor, and cryogenic plants; and related processing plants and terminals.

The code prescribes requirements for materials and components, design, fabrication, assembly, erection, examination, inspection, and testing of piping. This code applies to piping for all fluids, including (1) raw, intermediate, and finished chemicals; (2) petroleum products; (3) gas, steam, air, and water; (4) fluidized solids; (5) refrigerants; and (6) cryogenic fluids. Also included is piping that interconnects pieces or stages within a packaged equipment assembly.

B31.4, Pipeline Transportation Systems for Liquid Hydrocarbons and Other Liquids

This ASME code prescribes requirements for the design, materials, construction, assembly, inspection, and testing of piping transporting liquids such as crude oil, condensate, natural gasoline, natural gas liquids, liquefied petroleum gas, carbon dioxide, liquid alcohol, liquid anhydrous ammonia, and liquid petroleum products between producers' lease facilities, tank farms, natural gas processing plants, refineries, stations, ammonia plants, terminals (marine, rail, and truck), and other delivery and receiving points (see Figure 1–1).

Piping consists of pipe, flanges, bolting, gaskets, valves, relief devices, fittings, and the pressure-containing parts of other piping components. It also includes hangers and supports and other equipment items necessary to prevent overstressing the pressure-containing parts. It does not include support structures such as frames of buildings, buildings stanchions, or foundations. Requirements for offshore pipelines are found in Chapter IX of the code.

Also included within the scope of this code are (1) primary and associated auxiliary liquid petroleum and liquid anhydrous ammonia piping at pipeline terminals (marine, rail, and truck), tank farms, pump stations, pressure reducing stations, and metering stations, including scraper traps, strainers, and prover loops; (2) storage and working tanks, including pipe-type storage fabricated from pipe and fittings and the piping interconnecting these facilities; (3) liquid petroleum and liquid anhydrous ammonia piping located on property set aside for such piping within the petroleum refinery, natural gasoline, gas processing, ammonia, and bulk plants; (4) those aspects of operation and maintenance of liquid pipeline systems relating to the safety and protection of the general public, operating company personnel, environment, property, and the piping systems.

B31.5, Refrigeration Piping and Heat Transfer Components

This ASME code prescribes requirements for the materials, design, fabrication, assembly, erection, test, and inspection of refrigerant, heat transfer components, and secondary coolant piping for temperatures as low as –320°F (–196°C), whether erected on the premises or factory assembled, except as specifically excluded in the following paragraphs.

Figure 1–1 *Onshore pipelines are designed to ASME B31.4 for liquid transportation and ASME B31.8 for gas transmission. (Printed with the permission of Bentley Systems Incorporated.)*

Users are advised that other piping code sections may provide requirements for refrigeration piping in their respective jurisdictions. This code does not apply to (1) any self-contained or unit systems subject to the requirements of Underwriters Laboratories or other nationally recognized testing laboratory; (2) water piping; (3) piping designed for external or internal gauge pressure not exceeding 15 psig (105 kPa) regardless of size; or (4) pressure vessels, compressors, or pumps but does include all connecting refrigerant and secondary coolant piping starting at the first joint adjacent to such apparatus.

B31.8, Gas Transmission and Distribution Piping Systems

This ASME code covers the design, fabrication, installation, inspection, testing, and safety aspects of operation and maintenance of gas transmission and distribution systems, including gas pipelines, gas compressor stations, gas metering and regulation stations, gas mains, and service lines up to the outlet of the customer's meter set assembly (see Figure 1–1).

Included within the scope of this code are gas transmission and gathering pipelines, including appurtenances, that are installed offshore for the purpose of transporting gas from production facilities to onshore locations; gas storage equipment of the closed pipe type, fabricated or forged from pipe or fabricated from pipe and fittings; and gas storage lines.

B31.8S, Managing System Integrity of Gas Pipelines

This ASME standard applies to onshore pipeline systems constructed with ferrous materials that transport gas. *Pipeline system* means all parts of physical facilities through which gas is transported, including pipe, valves, appurtenances attached to pipe, compressor units, metering stations, regulator stations, delivery stations, holders, and fabricated assemblies.

The principles and processes embodied in integrity management are applicable to all pipeline systems. This standard is specifically designed to provide the operator (as defined in section 13) with the information necessary to develop and implement an effective integrity management program utilizing proven industry practices and processes. The processes and approaches within this standard are applicable to the entire pipeline system.

B31.9, Building Services Piping

This ASME code section has rules for the piping in industrial, institutional, commercial and public buildings, and multiunit residences, which does not require the range of sizes, pressures, and temperatures covered in B31.1. This code prescribes requirements for the design, materials, fabrication, installation, inspection, examination, and testing of piping systems for building services. It includes piping systems in the building or within the property limits.

B31.11, Slurry Transportation Piping Systems

This ASME code covers the design, construction, inspection, and security requirements of slurry piping systems. It covers piping systems that transport aqueous slurries of no hazardous materials, such as coal, mineral ores, and other solids, between a slurry processing plant and the receiving plant.

1.3 Codes 19

B31G, Manual for Determining Remaining Strength of Corroded Pipelines

This ASME manual includes all pipelines within the scope of the pipeline codes that are part of ASME B31, Code for Pressure Piping; that is, ASME B31.4, Liquid Transportation Systems for Hydrocarbons, Liquid Petroleum Gas, Anhydrous Ammonia, and Alcohols; ASME B31.8, Gas Transmission and Distribution Piping Systems; and ASME B31.11, Slurry Transportation Piping Systems. Parts 2, 3, and 4 are based on material included in the *ASME Guide for Gas Transmission and Distribution Piping Systems*, 1983 edition.

This manual is not applicable to new construction covered under the B31 code sections. That is, it is not intended that this manual be used to establish acceptance standards for pipe that may have become corroded prior to or during fabrication or installation. This manual is intended solely for the purpose of providing guideline information for the designer, owner, and operator. Therefore, the specific use of this manual is the responsibility of the designer, owner, or operator.

"The Design of Process Systems" concentrates on piping systems designed to the international code American Society of Mechanical Engineers B31.3, Process Piping. This document provides a minimum set of rules covering the design, materials, fabrication, examination, and testing of process piping systems, which includes but is not limited to petroleum refineries, oil and gas separation facilities, LNG plants, petrochemical complexes, and pharmaceutical plants.

By no means does this code cover all of the requirements for a process piping system, and it covers the broad spectrum of the subject and must be supplemented by other standards and specifications referenced within its pages.

B31.3, Process Piping

This ASME section covers the most commonly used code in the B31 series. It covers the design of chemical and petroleum plants and refineries processing chemicals and hydrocarbons, water, and steam. The code contains rules for piping typically found in petroleum refineries; chemical, pharmaceutical, textile, paper, semiconductor, and cryogenic plants; and related processing plants and terminals. ASME B31.3 also prescribes requirements for materials and components, design, fabrication, assembly, erection, examination, inspection, and testing of piping.

The code applies to piping for all fluids, including (1) raw, intermediate, and finished chemicals; (2) petroleum products; (3) gas, steam, air, and water, (4) fluidized solids, (5) refrigerants, and (6) cryogenic fluids. Also included is piping that interconnects pieces or stages within a packaged equipment assembly.

To maintain high-quality workmanship that will result in a safe environment and the use of standard materials, components, and methods of construction and testing, this code sets out a number of rules that cover

- Design.
- Material (strength).
- Flexibility (stress).
- Fabrication (welded joints).
- Erection (mechanical joints).
- Examination.
- Testing.

The basic function of ASME B31.3 code is to guarantee the safety of construction, commissioning, and operating personnel during the most critical periods of a plant's life. The various ASME design codes listed were introduced starting at the turn of the century, after a series of disasters that caused loss of life and major damage to facilities.

ASME B31.3, Process Piping, assumes that the plant life, that is, the length of time that it will be in operation, is 20 to 30 years, based on a safety factor of 3 to 1. These are for commercial projects. Plants that demand a very high level of reliability, because downtime has an immediate impact on power delivered to the general public are designed to ASME B31.1, Power Piping, which uses a safety factor of 4 to 1, which results in a plant life of approximately 40 years.

Between two facilities designed to ASME B31.3, the interconnecting pipelines are covered by ASME B31.4 for oil transportation and ASME

B31.8 for gas transportation. These two transportation codes are very similar in format to ASME B31.3 but have additional factors to consider, because a pipeline could be several thousands of miles long and travel through differing locations and encounter various environmental changes and climatic variations. In comparison, a process unit is in one geographical location that will be of several hundred acres but subjected to only one climate and, in a vast majority of cases, a common elevation, also known as *grade*.

It is very important for a piping engineer to study ASME 31.3 thoroughly and ideally invest in a personal copy. Remember that this code covers a wide spectrum of piping engineering—design, stress, materials, construction. If you are a specialist piping engineer, then there are sectors of this code you will rarely, if ever, use. Even so, it is important to know what the information is and who will use it. It is not essential to memorize the text and data, but it is very important to know what information is held within the pages of code and be able to access it and interpret it accurately. There are other national codes; however, none is as commonly used internationally as ASME B31.3 in the oil and gas industry.

The subjects covered within ASME B31.3 are

> Foreword.
>
> Committee Personnel.
>
> Introduction.
>
> Summary of Changes.
>
> Chapter I, Scope and Definitions.
>
> Chapter II, Design.
>
> > Part 1, Conditions and Criteria.
> >
> > Part 2, Pressure Design of Piping Components.
> >
> > Part 3, Fluid Service Requirements for Piping Components.
> >
> > Part 4, Fluid Service Requirements for Piping Joints.

Part 5, Flexibility and Support.

Part 6, Systems.

Chapter III, Materials.

Chapter IV, Standards for Piping Components.

Chapter V, Fabrication, Assembly, and Erection.

Chapter VI, Inspection, Examination, and Testing.

Chapter VII, Nonmetallic Piping and Piping Lined with Nonmetals.

Part 1, Conditions and Criteria.

Part 2, Pressure Design of Piping Components.

Part 3, Fluid Service Requirements for Piping Components.

Part 4, Fluid Service Requirements for Piping Joints.

Part 5, Flexibility and Support.

Part 6, Systems.

Part 7, Materials.

Part 8, Standards for Piping Components.

Part 9, Fabrication, Assembly, and Erection.

Part 10, Inspection, Examination, and Testing.

Chapter VIII, Piping for Category M Fluid Service.

Part 1, Conditions and Criteria.

Part 2, Pressure Design of Metallic Piping Components.

Part 3, Fluid Service Requirements for Metallic Piping Components.

Part 4, Fluid Service Requirements for Metallic Piping Joints.

Part 5, Flexibility and Support of Metallic Piping.

Part 6, Systems.

Part 7, Metallic Materials.

Part 8, Standards for Piping Components.

Part 9, Fabrication, Assembly, and Erection of Metallic Piping.

Part 10, Inspection, Examination, Testing, and Records of Metallic Piping.

Parts 11 through 20 correspond to Chapter VII.

Part 11, Conditions and Criteria.

Part 12, Pressure Design of Nonmetallic Piping Components.

Part 13, Fluid Service Requirements for Nonmetallic Piping Components.

Part 14, Fluid Service Requirements for Nonmetallic Piping Joints.

Part 15, Flexibility and Support of Nonmetallic Piping.

Part 16, Nonmetallic and Nonmetallic Lined Systems.

Part 17, Nonmetallic Materials.

Part 18, Standards for Nonmetallic and Nonmetallic-Lined Piping Components.

Part 19, Fabrication, Assembly, and Erection of Nonmetallic and Nonmetallic-Lined Piping.

Part 20, Inspection, Examination, Testing, and Records of Nonmetallic and Nonmetallic-Lined Piping.

Chapter IX, High Pressure Piping.

Part 1, Conditions and Criteria.

Part 2, Pressure Design of Piping Components.

Part 3, Fluid Service Requirements for Piping Components.

Part 4, Fluid Service Requirements for Piping Joints.

Part 5, Flexibility and Support.

Part 6, Systems.

Part 7, Materials.

Part 8, Standards for Piping Components.

Part 9, Fabrication, Assembly, and Erection.

Part 10, Inspection, Examination, and Testing.

Figures.

1.4 Standards and Specifications

ASME B31.3 is the design code supported by numerous standards and specifications that covers a great detail of information and data regarding the individual components that make up a piping system. Piping components are defined as pipe, pipe fittings, valves, gaskets, and bolts.

These international standards and specifications cover

- Materials—chemical composition, mechanical strength, and testing.
- Dimensions and tolerances.
- Examination.
- Testing of piping systems and valves.
- Fabrication.

The most commonly used standards and specifications under the umbrella of the ASME B31.3 code are from the following organizations:

- American Society of Mechanical Engineers (ASME).
- American Petroleum Institute (API).
- Manufacturers Standardization Society (MSS).
- American Society for Testing and Materials (ASTM).
- American Water Works Association (AWWA).
- American Welding Society (AWS).
- American Society for Nondestructive Testing (ASNT).
- American Iron and Steel Institute (AISI).
- Manufacturers Standardization Society of the Valves and Fittings Industry (MSS).
- National Association of Corrosion Engineers (NACE).
- National Fire Protection Association (NFPA).
- Pipe Fabrication Institute (PFI).
- Society of Automotive Engineers (SAE).

A list of standards and specifications referenced in ASME B31.3 can be found in the next section.

The purpose for the publication of these standards and specifications is to standardize the dimensions, tolerances, and the materials of construction of piping components to make the fabrication and erection less complex. Procedures and inspection standards and specifications ensure that the quality of the material and the completion of a piping system are to a level acceptable to the relevant code. These documents are essential, and without them, the industry would be anarchic.

1.4.1 American Society of Mechanical Engineers

The series of ASME standards that follow are primarily dimensional standards for piping components:

B1.1, Standard for Screw Threads.

B1.20.1, Pipe Threads, General Purpose, Inch.

B16.1, Cast Iron Pipe Flanges and Flanged Fittings.

B16.3, Malleable Iron Threaded Fittings.

B16.4, Cast Iron Threaded Fittings.

B16.5, Pipe Flanges and Flanged Fittings.

B16.9, Factory-Made Wrought Steel Butt Welding Fittings.

B16.10, Face-to-Face and End-to-End Dimensions of Valves.

B16.11, Forged Steel Fittings, Socket-Welding and Threaded.

B16.14, Ferrous Pipe Plugs, Bushings and Locknuts with Pipe Threads.

B16.15, Cast Bronze Threaded Fittings.

B16.18, Cast Copper Alloy Solder Joint Pressure Fittings.

B16.20, Metallic Gaskets for Pipe Flanges—Ring Joint, Spiral-Wound, and Jacketed.

B16.21, Nonmetallic Flat Gaskets for Pipe Flanges.

B16.22, Wrought Copper and Copper Alloy Solder Joint Pressure Fittings.

B16.24, Cast Copper Alloy Pipe Flanges and Flanged Fittings.

B16.25, Butt Welding Ends.

B16.26, Cast Copper Alloy Fittings for Flared Copper Tubes.

B16.28, Wrought Steel, Butt Welding, Short Radius Elbows and Returns.

B16.34, Valves—Flanged, Threaded, and Welding End.

16.36, Orifice Flanges.

B16.39, Malleable Iron Threaded Pipe Unions.

B16.42, Ductile Iron Pipe Flanges and Flanged Fittings, Classes 150 and 300.

B16.47, Large Diameter Steel Flanges: NPS 26 through NPS 60.

B16.48, Steel Line Blanks.

B36.10M, Welded and Seamless Wrought Steel Pipe.

B36.19M, Stainless Steel Pipe.

Next, we discuss the scopes of most commonly used ASME standards that include mechanical and dimensional data, which allows the standardization of piping systems and is beneficial for both design and construction personnel.

B1.20.1, 1983, Pipe Threads, General Purpose, Inch (Scope)

This ANSI standard covers the dimensions and gauging of pipe threads for general purpose applications. The B1.20.1 code is a revision and redesignation of ANSI B2.1,1968. The inclusion of dimensional data in this standard is not intended to imply that all the products described are stock production sizes. Consumers must consult with manufacturers concerning availability of products. Metric, general purpose, semitubular rivets purchased for government use conform to this standard and, additionally, to the requirements of Appendix I.

B16.5, 2003, Pipe Flanges and Flanged Fittings: NPS ½ through 24 (Scope)

This standard covers pressure-temperature ratings, materials, dimensions, tolerances, marking, testing, and methods of designating openings for pipe flanges and flanged fittings. Included are flanges with rating class designations 150, 300, 400, 600, 900, 1500, and 2500 in sizes NPS ½ through NPS 24, with requirements given in both metric and U.S. customary units, with diameter of bolts and flange bolt holes expressed in inch units; flanged fittings with rating class designation

150 and 300 in sizes NPS ½ through NPS 24, with requirements given in both metric and U.S. customary units, with diameter of bolts and flange bolt holes expressed in inch units; and flanged fittings with rating class designation 400, 600, 900, 1500, and 2500 in sizes NPS ½ through NPS 24 that are acknowledged in Annex G, in which only U.S. customary units are provided. This standard is limited to flanges and flanged fittings made from cast or forged materials, blind flanges, and certain reducing flanges made from cast, forged, or plate materials. Also included in this standard are requirements and recommendations regarding flange bolting, flange gaskets, and flange joints.

B16.9, 2003, Factory-Made Wrought Butt Welding Fittings (Scope)

This standard covers overall dimensions, tolerances, ratings, testing, and markings for wrought carbon and alloy steel, factory-made, butt-welded fittings of NPS ½ through 48. It covers fittings of any producible wall thickness. This standard does not cover low-pressure, corrosion-resistant, butt-welding fittings. See MSS SP-43, Wrought Stainless Steel Butt-Welded Fittings.

B16.10, 2000, Face-to-Face and End-to-End Dimensions of Valves

This standard covers face-to-face and end-to-end dimensions of straightway valves and center-to-face and center-to-end dimensions of angle valves. Its purpose is to assure installation interchangeability for valves of a given material, type, size, rating class, and end connection.

B16.11, 1996, Forged Fittings, Socket Welding and Threaded (Scope)

This standard covers ratings, dimensions, tolerances, marking, and material requirements for forged fittings, both socket welded and threaded.

B16.20, 1998, Metallic Gaskets for Pipe Flanges: Ring Joint Spiral Wound and Jacketed (Scope)

This standard covers materials, dimensions, tolerances, and markings for metal ring-joint gaskets, spiral-wound metal gaskets, and metal-jacketed gaskets and filler material. These gaskets are dimensionally suitable for use with the flanges described in the reference flange standards ASME B16.5, ASME B16.47, and API-6A. This standard covers spiral-wound metal gaskets and metal-jacketed gaskets for use with raised face and flat face flanges.

B16.21, 2005, Nonmetallic Flat Gaskets for Pipe Flanges (Scope)

This standard covers types, sizes, materials, dimensions, tolerances, and markings for nonmetallic flat gaskets. These gaskets are dimensionally suitable for use with flanges described in the referenced flange standards.

B16.34, 2004, Valves Flanged, Threaded and Welding End (Scope)

This standard applies to new construction and covers pressure-temperature ratings, dimensions, tolerances, materials, nondestructive examination requirements, testing, and marking for cast, forged, and fabricated flanged, threaded, and welded end and wafer or flangeless valves of steel, nickel-base alloys and other alloys shown in Table 1 of the ASME standards. Wafer or flangeless valves, bolted or through-bolt types, which are installed between flanges or against a flange, are treated as flanged-end valves. Alternative rules for NPS 2½ and smaller valves are given in Mandatory Appendix V.

B16.36, 1996, Orifice Flanges (Scope)

This standard covers flanges (similar to those covered in ASME B16.5) that have orifice pressure differential connections. Coverage is limited to the following: (1) welding neck flanges classes 300, 400, 600, 900, 1500, and 2500 and (2) slip-on and threaded class 300.

B16.39, 1998, Malleable Iron Threaded Pipe Unions (Scope)

This standard for threaded malleable iron unions, classes 150, 250, and 300, provides requirements for the following: (1) design, (2) pressure-temperature ratings, (3) size, (4) marking, (5) materials, (6) joints and seats, (7) threads, (8) hydrostatic strength, (9) tensile strength, (10) air pressure test, (11) sampling, (12) coatings, and (13) dimensions.

B16.47, 1996, Large Diameter Steel Flanges (Scope)

This standard covers pressure-temperature ratings, materials, dimensions, tolerances, marking, and testing for pipe flanges in sizes NPS 26 through NPS 60 and ratings classes 75, 150, 300, 400, 600, and 900. Flanges may be of cast, forged, or plate (for blind flanges only) materials, as listed in Table 1A. Requirements and recommendations regarding bolting and gaskets are included.

B16.48, 1997, Steel Line Blanks (Scope)

This standard covers pressure-temperature ratings, materials, dimensions, tolerances, marking, and testing for operating line blanks in sizes NPS ½ through NPS 24 for installation between ASME B16.5 flanges in the 150, 300, 600, 900, 1500, and 2500 pressure classes. The dimensions are suitable for blanks made of materials listed in Table 1.

B36.10M, 2004, Welded and Seamless Wrought Steel Pipe (Scope)

This standard covers the standardization of dimensions of welded and seamless wrought steel pipe for high or low temperatures and pressures. The word *pipe* is used as distinguished from tube to apply to tubular products of dimensions commonly used for pipeline and piping systems. Pipe NPS 12 (DN 300) and smaller have outside diameters numerically larger than corresponding sizes. In contrast, the outside diameters of tubes are numerically identical to the size number for all sizes.

B36.19M, 1985, Stainless Steel Pipe (Scope)

This standard covers the standardization of dimensions of welded and seamless wrought stainless steel pipe. The word *pipe* is used as distinguished from tube to apply to tubular products of dimensions commonly used for pipeline and piping systems. Pipe dimensions of sizes 12 and smaller have outside diameters numerically larger than the corresponding size. In contrast, the outside diameters of tubes are numerically identical to the size number for all sizes. The wall thicknesses for sizes 14 through 22 inclusive of schedule 10S, for size 12 of schedule 40S, and for sizes 10 and 12 of schedule 80S are not the same as those of ANSI/ASME B36.10M. The suffix S in the schedule number is used to differentiate B36.19M pipe from B36.10M pipe. ANSI/ASME B36.10M includes other pipe thicknesses, which are also commercially available in stainless steel material.

1.4.2 American Petroleum Institute

The American Petroleum Institute publishes specifications (Spec.), bulletins (Bull.), recommended practices (RP), standards (Std.), and publications (Publ.) as an aid to procurement of standardized equipment and materials for the petroleum industry.

The following documents, which relate to piping, are published by the API:

Specifications (Spec.)

Spec. 5B

Spec. 5L, Specification for Line Pipe.

Spec. 15LE, Specification for Polyethylene Line Pipe (PE).

Spec. 15LR, Specification for Low Pressure Fiberglass Line Pipe.

Recommended Practices (RP)

RP 941.

Standards (Std.)

Std. 526, Flanged Steel Pressure-Relief Valves Seat.

Std. 594, Wafer and Wafer-Lug Check Valves.

Std. 599, Metal Plug Valves—Flanged and Welding Ends.

Std. 600, Steel Gate Valves—Flanged and Butt- Welding Ends.

Std. 602, Compact Steel Gate Valves—Flanged, Threaded, Welding, and Extended Body Ends.

Std. 603, Class 150, Cast, Corrosion-Resistant, Flanged-End Gate Valves.

Std. 608, Metal Ball Valves—Flanged, Threaded, and Welding Ends.

Std. 609, Lug and Wafer Type Butterfly Valves.

Publications (Publ.)

Publ. 1113, Developing a Pipeline Supervisory Control Center.

1.4.3 American Society for Testing and Materials

The American Society for Testing and Materials (ASTM) is a scientific and technical organization that develops and publishes voluntary standards on the characteristics and performance of materials, products, standards, systems, and services. The standards published by the ASTM include test procedures for determining or verifying characteristics, such as chemical composition, and measuring performance, such as tensile strength and bending properties. The standards cover refined materials, such as steel, and basic products, such as machinery and fabricated equipment. The standards are developed by committees drawn from a broad spectrum of professional, industrial, and commercial interests. Many of the standards are made mandatory by reference in applicable piping codes. The ASTM standards are published in a set of 67 volumes divided into 16 sections. Each volume is published annually to incorporate new standards and revisions to existing standards and to delete obsolete standards.

Listed here are the 67 volumes, divided among their 16 sections, published by the ASTM:

Section 1, Iron and Steel Products.

Volume 01.01, Steel—Piping, Tubing, Fittings.

Volume 01.02, Ferrous Castings; Ferroalloys.

Volume 01.03, Steel—Plate, Sheet, Strip, Wire.

Volume 01.04, Steel—Structural, Reinforcing, Pressure Vessel, Railway.

Volume 01.05, Steel—Bars, Forgings, Bearing, Chain, Springs.

Volume 01.06, Coated Steel Products.

Volume 01.07, Shipbuilding.

Section 2, Nonferrous Metal Products.

Volume 02.01, Copper and Copper Alloys.

Volume 02.02, Aluminum and Magnesium Alloys.

Volume 02.03, Electrical Conductors.

Volume 02.04, Nonferrous Metals—Nickel, Cobalt, Lead, Tin, Zinc, Cadmium, Precious, Reactive, Refractory, Metals, and Alloys.

Volume 02.05, Metallic and Inorganic Coatings; Metal Powders, Sintered P/M Structural Parts.

Section 3, Metals Test Methods and Analytical Procedures.

Volume 03.01, Metals—Mechanical Testing: Elevated and Low-Temperature Tests, Metallography.

Volume 03.02, Wear and Erosion, Metal Corrosion.

Volume 03.03, Nondestructive Testing.

Volume 03.04, Magnetic Properties; Metallic Materials for Thermostats, Electrical Heating and Resistance, Heating, Contacts, and Connectors.

Volume 03.05, Analytical Chemistry of Metals, Ores, and Related Materials (I).

Volume 03.06, Analytical Chemistry of Metals, Ores, and Related Materials (II).

Section 4, Construction.

Volume 04.01, Cement, Lime, Gypsum.

Volume 04.02, Concrete and Aggregates.

Volume 04.03, Road and Paving Materials, Pavement Management Technologies.

Volume 04.04, Roofing, Waterproofing, and Bituminous Materials.

Volume 04.05, Chemical-Resistant Materials; Vitrified Clay, Concrete, Fiber-Cement Products; Mortars; Masonry.

Volume 04.06, Thermal Insulation; Environmental Acoustics.

Volume 04.07, Building Seals and Sealants; Fire Standards; Building Constructions.

Volume 04.08, Soil and Rock; Dimension Stones; Geosynthetics.

Volume 04.09, Wood.

Section 5, Petroleum Products, Lubricants, and Fossil Fuels.

Volume 05.01, Petroleum Products and Lubricants (1): D 56-D 1947.

Volume 05.02, Petroleum Products and Lubricants (II): D 1949-D 3601.

Volume 05.03, Petroleum Products and Lubricants (III): D 3602–Latest; Catalysts.

Volume 05.04, Test Methods for Rating Motor, Diesel, and Aviation Fuels.

Volume 05.05, Gaseous Fuels; Coal and Coke.

Section 6, Paints, Related Coatings, and Aromatics.

Volume 06.01, Paint—Tests for Formulated Products and Applied Coatings.

Volume 06.02, Paint—Pigments, Resins, and Polymers; Cellulose.

Volume 06.03, Paint—Fatty Oils and Acids, Solvents, Miscellaneous; Aromatic Hydrocarbons.

Section 7, Textiles.

Volume 07.01, Textiles (I): D76-D3219.

Volume 07.02, Textiles (II): D3333–Latest.

Section 8, Plastics.

Volume 08.01, Plastics (I): C 177-D 1600.

Volume 08.02, Plastics (II): D 1601-D 3099.

Volume 08.03, Plastics (III): D 3100–Latest.

Volume 08.04, Plastic Pipe and Building Products.

Section 9, Rubber.

Volume 09.01, Rubber, Natural, and Synthetic—General Test Methods; Carbon Black.

Volume 09.02, Rubber Products, Industrial—Specifications and Related Test Methods; Gaskets; Tires.

Section 10, Electrical Insulation and Electronics.

Volume 10.01, Electrical Insulation (I) D69-D2484.

Volume 10.02, Electrical Insulation (II) D2518–Latest.

Volume 10.03, Electrical Insulating Liquids and Gas; Electrical Protective Equipment.

Volume 10.04, Electronics (I).

Volume 10.05, Electronics (II).

Section 11, Water and Environmental Technology.

Volume 11.01, Water (I).

Volume 11.02, Water (II).

Volume 11.03, Atmospheric Analysis; Occupational Health and Safety.

Volume 11.04, Pesticides; Resource Recovery; Hazardous Substances and Oil Spill. Responses; Waste Management; Biological Effects.

Section 12, Nuclear, Solar, and Geothermal Energy.

Volume 12.01, Nuclear Energy (I).

Volume 12.02, Nuclear, Solar, and Geothermal Energy.

Section 13, Medical Devices and Services.

> Volume 13.01, Medical Devices, Emergency Medical Services.

Section 14, General Methods and Instrumentation.

> Volume 14.01, Analytical Methods—Spectroscopy; Chromatography; Computerized Systems.
>
> Volume 14.02, General Test Methods, Nonmetal; Laboratory Apparatus; Statistical Methods; Appearance of Materials; Durability of Nonmetallic Materials.
>
> Volume 14.03, Temperature Measurement.

Section 15, General Products, Chemical Specialties, and End-Use Products.

> Volume 15.01, Refractures; Carbon and Graphite Products; Activated Carbon.
>
> Volume 15.02, Glass; Ceramic Whitewares.
>
> Volume 15.03, Space Simulation; Aerospace and Aircraft; High Modulus Fibers and Composites.
>
> Volume 15.04, Soap; Polishes; Leather; Resilient Floor Coverings.
>
> Volume 15.05, Engine Coolants; Halogenated Organic Solvents; Industrial Chemicals.
>
> Volume 15.06, Adhesives.
>
> Volume 15.07, End-Use Products.
>
> Volume 15.08, Fasteners.
>
> Volume 15.09, Paper; Packaging; Flexible Barrier Materials; Business Copy Products.

Section 00, Index.

> Volume 00.01, Subject Index and Alphanumeric List.

A typical American Standard for Testing Materials covers, among many things, the following;

- Scope.
- Referenced documents.
- Ordering information.
- Chemical composition.
- Mechanical requirements—tensile strength, yield strength, elongation.
- Heat treatment.
- Workmanship, finish, and appearance.
- Additional supplementary requirements.

These standards guarantee that, if the base material used is manufactured to a set of well defined rules, then the various characteristics of that material will be predictable and the materials of construction can be specified with confidence.

The following is an example of the detail that an ASTM specification goes into:

> Section 1, Iron and Steel Products.
>
> Volume 01.01, Steel—Piping, Tubing, Fittings.
>
> ASTM A 106/A 106M–04b, Standard Specification for Seamless Carbon Steel Pipe for High-Temperature Service.
>
> (1) Scope.
>
> (2) Referenced documents.
>
> (3) Ordering information.
>
> (4) Process.
>
> (5) Heat treatment.

(6) General requirements.

(7) Chemical composition.

(8) Heat analysis.

(9) Product analysis.

(10) Tensile requirements.

(11) Bending requirements.

(12) Flattening tests.

(13) Hydrostatic test.

(14) Nondestructive electric test.

(15) Nipples.

(16) Dimensions, mass, and permissible variations.

(17) Lengths.

(18) Workmanship, finish, and appearance.

(19) End finish.

(20) Number of tests.

(21) Retests.

(22) Test specimens and test methods.

(23) Certification.

(24) Product marking.

(25) Government procurement.

(26) Keywords.

Supplementary Requirements

(S1) Product analysis.

(S2) Transverse tension test.

(S3) Flattening test.

(S4) Metal structure and etching test.

(S5) Carbon equivalent.

(S6) Heat treated test specimens.

(S7) Internal cleanliness—government orders.

(S8) Requirements for carbon steel pipe for hydrofluoric acid alkylation service.

1.4.4 American Society for Nondestructive Testing

The American Society for Nondestructive Testing (ASNT) publishes recommended practices concerning procedures, equipment, and qualifications of personnel for nondestructive testing. The following practice is cited in several codes and standards that contain requirements for piping:

SNT-TC-1A, Recommended Practice for Nondestructive Testing Personnel Qualifications.

1.4.5 American Society for Quality

The following are some of the American Society for Quality (ASQ) standards:

Q9000-1, 1994, Quality Management and Quality Assurance Standards—Guidelines for Selection and Use.

Q9000-2, 1997, E-Standard.

Q9000, 3-1997, E-Standard.

Q 9001, 2000, Excerpted Version: Quality Management System—Requirements: An Abridged Document for Auditors (E-Standard).

Q9002, 1994, Quality Systems—Model for Quality Assurance in Production, Installation, and Servicing.

Q9003, 1994, Model for Quality Assurance in Final Inspection and Test—E-Standard.

1.4.6 American Welding Society

The American Welding Society (AWS) publishes handbooks, manuals, guides, recommended practices, specifications, and codes. The specifications for filler metals are in the AWS A5 series. The filler metal specifications usually are cited in design documents. The welding procedures are in the D10 series. The AWS handbook is published in five volumes and intended to be an aid to the user and producer of welded products.

The following is a list of AWS publications directly related to piping:

> AWS Welding Handbook
>
>> Volume 1, Fundamentals of Welding.
>>
>> Volume 2, Welding Processes.
>>
>> Volume 3, Welding Processes.
>>
>> Volume 4, Engineering Applications—Materials.
>>
>> Volume 5, Engineering Applications—Design Brazing Manual, Soldering Manual.
>
> AWS A3.0, Soldering Manual, Brazing Handbook, Welding Terms and Definitions, Including Terms for Brazing, Soldering, Thermal Spraying, and Thermal Cutting.
>
> AWS A5.01, Filler Metal Procurement Guidelines.
>
> AWS D10.4, Recommended Practices for Welding Austenitic Chromium-Nickel Stainless Steel Piping and Tubing.
>
> AWS D10.6, Recommended Practices for Gas Tungsten Arc Welding of Titanium Pipe and Tubing.
>
> AWS D10.7, Recommended Practices for Gas Shielded Arc Welding of Aluminum and Aluminum Alloy Pipe.
>
> AWS D10.8, Recommended Practices for Welding of Chromium-Molybdenum Steel Piping and Tubing.
>
> AWS D10.10, Recommended Practices for Local Heating of Welds in Piping and Tubing.

AWS D10.11, Recommended Practices for Root Pass Welding of Pipe without Backing.

AWS D10.12, Recommended Practices for Procedures for Welding Low Carbon Steel Pipe.

1.4.7 American Water Works Association

The American Water Works Association (AWWA) publishes standards that cover requirements for pipe and piping components used in water treatment and distribution systems, including specialty items, such as fire hydrants.

The AWWA standards are used for the design, fabrication, and installation of large-diameter piping for water systems not covered by the ASME Boiler and Pressure Vessel Code, ASME B31, Code for Pressure Piping, and other codes. Conformance to AWWA standards is required either by being referenced in the codes governing the construction of water systems piping or by the enforcement authorities having jurisdiction over the water systems piping.

Ductile-Iron Pipe and Fittings

C110/A21.10, ANSI Standard for Ductile-Iron and Gray-Iron Fittings, 3–48" (76–1219 mm), for Water.

C111/A21.11. ANSI Standard for Rubber-Gasket Joints for Ductile-Iron Pressure Pipe and Fittings.

C115/A21.15, ANSI Standard for Flanged Ductile-Iron Pipe with Ductile-Iron or Gray-Iron Threaded Flanges.

C150/A21.50, ANSI Standard for Thickness Design of Ductile-Iron Pipe.

C151/A21.51, ANSI Standard for Ductile-Iron Pipe, Centrifugally Cast, for Water.

Steel Pipe

C200, Steel Water Pipe—6" (150 mm) and Larger.

C207, Steel Pipe Flanges for Waterworks Service—Sizes 4" through 144" (100 mm through 3600 mm).

C208, Dimensions for Fabricated Steel Water Pipe Fittings.

Concrete Pipe

C300, Reinforced Concrete Pressure Pipe, Steel-Cylinder Type.

C301, Prestressed Concrete Pressure Pipe, Steel-Cylinder Type.

C302, Reinforced Concrete Pressure Pipe, No-Cylinder Type.

Valves and Hydrants

C500, Metal-Seated Gate Valves for Water Supply Service (Includes Addendum C500a-95).

C504, Rubber-Seated Butterfly Valves.

Plastic Pipe

C900-97, Polyvinyl Chloride (PVC) Pressure Pipe, and Fabricated Fittings, 4–12" (100–300 mm), for Water Distribution.

C950-01, Fiberglass Pressure Pipe.

1.4.8 Copper Development Association

The Copper Development Association (CDA) is the market development, engineering, and information services arm of the copper industry, chartered to enhance and expand markets for copper and its alloys in North America. The relevant CDA publication is the *Copper Tube Handbook*, 1995.

1.4.9 Compressed Gas Association

The Compressed Gas Association (CGA) promotes the safe manufacture, transportation, storage, transfilling, and disposal of industrial and medical gases and their containers. The relevant CGA publication is G-4.1, 1996, *Cleaning Equipment for Oxygen Service*, 4th edition.

1.4.10 Canadian Standards Association

The Canadian Standards Association (CSA) is a not-for-profit, membership-based association serving business, industry, government, and consumers in Canada and the global marketplace. The CSA works in Canada and around the world to develop standards that address real needs, such as enhancing public safety and health. The relevant CSA publication is Z245.1, 1998, *Steel Pipe*.

1.4.11 Expansion Joint Manufacturers Association

The Expansion Joint Manufacturers Association (EJMA) represents established manufacturers of metal bellows-type expansion joints. EJMA was founded in 1955 to establish and maintain quality design and manufacturing standards. The relevant EJMA publication is *EJMA Standards*, 8th edition, 2003.

1.4.12 Manufacturers Standardization Society of the Valve and Fittings Industry

The Manufacturers Standardization Society (MSS) publishes standard practices (SP), which provide a basis for common practice by the manufacturers, the user, and the general public. Compliance to the standard practices of MSS is required by reference in a code, specification, sales contract, law, or regulation. The MSS is also represented on the committees of other standardization groups, such as ASME.

The following is a complete list of MSS standard practices published and in current use:

> SP-6, Standard Finishes for Contact Faces of Pipe Flanges and Connecting-End Flanges of Valves and Fittings.
>
> SP-9, Spot Facing for Bronze, Iron, and Steel Flanges.

SP-25, Standard Marking System for Valves, Fittings, Flanges, and Unions.

SP-42, Class 150 Corrosion Resistant Gate, Globe, Angle, and Check Valves with Flanged and Butt-Weld Ends.

SP-43, Wrought Stainless Steel Butt-Welding Fittings, Including Reference to Other Corrosion Resistant Materials.

SP-44, Steel Pipe Line Flanges (superseded by ASME B16.47).

SP-45, Bypass and Drain Connection Standard.

SP-51, Class 15OLW Corrosion Resistant Cast Flanges and Flanged Fittings.

SP-53, Quality Standard for Steel Castings and Forgings for Valves, Flanges, and Fittings and Other Piping Components Magnetic Particle Examination Method.

SP-55, Quality Standard for Steel Castings for Valves, Flanges, and Fittings and Other Piping Components, Visual Method for Evaluation of Irregularities.

SP-58, Pipe Hangers and Supports—Materials, Design, and Manufacture.

SP-65, High Pressure Chemical Industry Flanges and Threaded Stubs for Use with Lens Gaskets.

SP-70, Cast Iron Gate Valves, Flanged, and Threaded Ends.

SP-71, Cast Iron Swing Check Valves, Flanged and Threaded Ends.

SP-72, Ball Valves with Flanged or Butt-Welding Ends for General Service.

SP-73, Brazing Joints for Wrought and Cast Copper Alloy Solder Joint Pressure Fittings.

SP-75, Specification for High Test Wrought Butt-Welding Fittings.

SP-79, Socket-Welding Reducer Inserts.

SP-80, Bronze Gate, Globe, Angle, and Check Valves.

SP-81, Stainless Steel, Bonnetless Flanged Knife Gate Valves.

SP-83, Class 3000 Steel Pipe Unions Socket-Welding and Threaded.

SP-85, Cast Iron Globe and Angle Valves Flanged and Threaded Ends.

SP-88, Diaphragm Type Valves (R 1988).

SP-97, Forged Carbon Steel Branch Outlet Fittings— Socket Welding, Threaded, and Butt-Welding Ends.

MSS SP-119, Belled End Socket Welding Fittings, Stainless Steel and Copper Nickel.

1.4.13 National Association of Corrosion Engineers

The National Association of Corrosion Engineers (NACE) mission to protect people, assets, and the environment from the effects of corrosion is global. Its members work in corporations, educational institutions, research facilities, and federal governments. The relevant NACE publications are

MR0175, Metals for Sulfide Stress Cracking and Stress Corrosion Cracking Resistance in Sour Oilfield Environments.

RP0170, Protection of Austenitic Stainless Steels and Other Austenitic Alloys from Polythionic Acid Stress Corrosion Cracking During Shutdown of Refinery Equipment. RP0472, Methods and Controls to Prevent In-Service Environmental Cracking of Carbon Steel Weldments in Corrosive Petroleum Refining Environments.

1.4.14 National Fire Protection Association

The National Fire Protection Association (NFPA) represents all aspects of fire protection, such as professional societies, educational institutions, public officials, insurance companies, equipment manufacturers, builders and contractors, and transportation groups. The NFPA publishes codes, standards, guides, and recommended practices in a 12-volume set of books called the *National Fire Codes*.

Volumes 1 through 8 contain actual text of the national fire codes and standards. The requirements contained in these volumes have been judged suitable for legal adoption and enforcement. Volumes 9 through 11 contain recommended practices and guides considered to be good engineering practices. Volume 12 contains formal interpretations, tentative interim amendments, and errata that relate to the documents in Volumes 1 through 11.

1.4.15 Pipe Fabrication Institute

The Pipe Fabrication Institute (PFI) publishes advisory engineering standards (ES) and technical bulletins (TB) intended to serve the needs of the pipe-fabricating industry at the design level and in actual shop operations. The PFI standards contain minimum requirements; however, the designer or fabricator may consider specifying additional requirements beyond the scope of PFI publications. The use of PFI standards or bulletins is voluntary. A complete listing of PFI publications follows:

> ES-1, Internal Machining and Solid Machined Backing Rings for Circumferential Butt Welds.
>
> ES-2, Method of Dimensioning Piping Assemblies.
>
> ES-3, Fabricating Tolerances.
>
> ES-4, Hydrostatic Testing of Fabricated Piping.
>
> ES-5, Cleaning of Fabricated Piping.
>
> ES-7, Minimum Length and Spacing for Welded Nozzles.
>
> ES-11, Permanent Marking on Piping Materials.
>
> ES-16, Access Holes, Bosses, and Plugs for Radiographic Inspection of Pipe Welds ES-20 Wall Thickness Measurement by Ultrasonic Examination.
>
> ES-21, Internal Machining and Fit-Up of GTAW Root Pass Circumferential Butt Welds.
>
> ES-22, Recommended Practice for Color Coding of Piping Materials.

ES-24, Pipe Bending Methods, Tolerances, Process, and Material Requirements.

ES-25, Random Radiography of Pressure Retaining Girth Butt Welds.

ES-26, Welded Load Bearing Attachments to Pressure Retaining Piping Materials.

ES-27, "Visual Examination"—The Purpose, Meaning, and Limitation of the Term.

ES-29, Abrasive Blast Cleaning of Ferritic Piping Materials.

ES-30, Random Ultrasonic Examination of Butt Welds.

ES-31, Standard for Protection of Ends of Fabricated Piping Assemblies.

ES-32, Tool Calibration.

ES-34, Painting of Fabricated Piping.

ES-35, Nonsymmetrical Bevels and Joint Configurations for Butt Welds.

ES-36, Branch Reinforcement Work Sheets.

PFI ES-37, Loading and Shipping of Piping Assemblies.

PFI ES-39, Fabricated Tolerances for Grooved Piping Systems.

PFI ES-40, Method of Dimensioning Grooved Piping Assemblies.

PFI ES-41, Material Control and Traceability of Piping Components.

PFI ES-42, Positive Material Identification of Piping Components Using Portable X-Ray Emission-Type Test Equipment.

PFI ES-44, Drafting Practices Standard.

TB1, Pressure-Temperature Ratings of Seamless Pipe Used in Power Plant Piping Systems.

TB3, Guidelines Clarifying Relationships and Design Engineering Responsibilities between Purchasers' Engineers and Pipe Fabricator or Pipe Fabricator Erector.

TB7, Guideline for Fabrication and Installation of Stainless Steel High Priority Distribution Systems.

1.4.16 Society of Automotive Engineers

The Society of Automotive Engineers (SAE) is an international organization for mobility engineering professionals in the aerospace, automotive, and commercial vehicle industries. The society is a standards development organization for the engineering of powered vehicles of all kinds, including cars, trucks, boats, aircraft, and others. Its standards also are used outside these industries. The relevant SAE publications are

SAE J513, Refrigeration Tube Fittings—General Specifications.

SAE J514, Hydraulic Tube Fittings.

SAE J518, Hydraulic Flanged Tube, Pipe, and Hose Connections, Four-Bolt Split Flange Type.

CHAPTER 2

Piping Components

Another factor that remains unchanged, in spite of the CAD technological advances and the digital representation of piping systems, is the piping components.

This chapter deals with the numerous types of piping components that make up a process piping system. The selection of the design and the materials of construction is extremely important and should be based on the past performance of the piping component in similar or more extreme design conditions. Rarely will a piping engineer or designer be faced with selection decisions that have not occurred on a previous project somewhere in the world. It is essential that the individual is fully aware of the limitations of the component and all of the design conditions. The chapter is divided into the following sections:

2.1. Introduction to Piping Components

2.2. Pipe

2.3. Piping Fittings

2.4. Flanges

2.5. Valves

2.6. Bolts and Gaskets (Fasteners and Sealing)

2.1 Introduction to Piping Components

To connect the various process and utility equipment contained within a process plant, it is necessary to use an assortment of piping components that, when used collectively, are called a *piping system*. This chapter introduces the reader to these components and explains their design function and how they are specified, manufactured, and installed. All components have their own characteristics, both positive and negative, and it is essential to be aware of their strengths and weaknesses. Specifying them can become complex, especially for valves and piping special items.

The individual components necessary to complete a piping system are

- Pipe.
- Piping fittings.
- Valves.
- Bolts and gaskets (fasteners and sealing).
- Piping special items, such as steam traps, pipe supports, and valve interlocking.

These pressure-containing and non-pressure-containing components combine to form the ingredients of a piping system.

I introduce each category and outline the general international standards and specifications that apply to that particular group of components. Although individual components have different commercial values and availability, all are of equal importance in a piping system that is to function safely and efficiently (see Figure 2–1).

For example, you could have a very expensive valve held in position by two, comparatively less expensive flanges, two gaskets, and a set of bolting worth a fraction of the cost but no less important. The specification and the correct installation procedure of mating flanges, gaskets, and bolts are essential for the valve to be installed and function efficiently within a piping system.

I start with the least complex component within a process piping system.

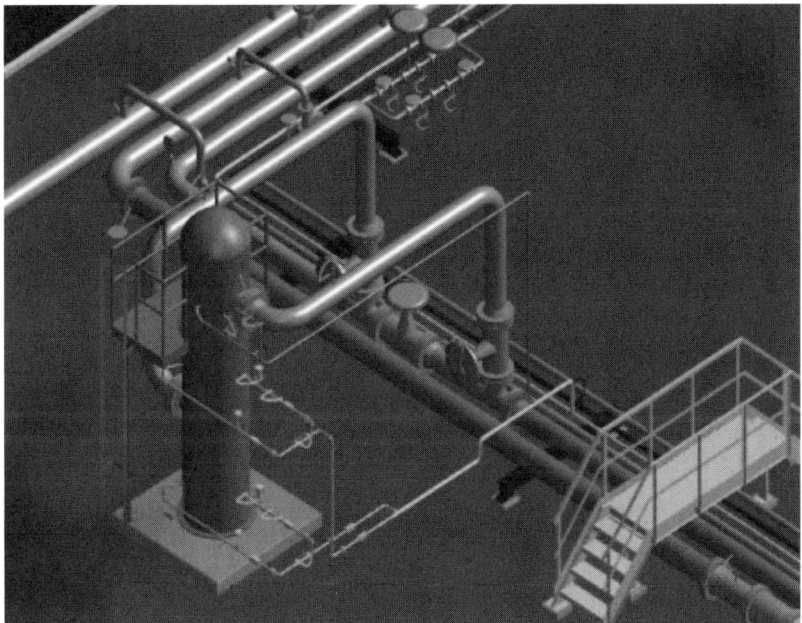

Figure 2–1 *A 3D model that shows a variety of components used to pipe up a vertical vessel. (Printed with the permission of Bentley Systems Incorporated.)*

2.2 Pipe

Pipe is the main artery that connects the various pieces of process and utility equipment within a process plant. Although it can be considered to be the least complex component within a piping system, it is not without its peculiarities. Pipe used within a process plant designed to one of the ASME B31 codes generally is of a metallic construction, such as carbon steel, stainless steel, duplex, copper, or to a lesser degree, one of the more exotic metals like Monel or titanium.

Nonmetallic pipe such as one of the plastics, like PVC, glass-reinforced epoxy, or glass-reinforced plastic, are not prohibited, and each has its own set of characteristics. Glass-reinforced plastic (GRP), is a plastic reinforced by fine fibers of glass. The plastic most commonly used is polyester or vinylester, but other plastics, such as epoxy, can be used to make glass-reinforced epoxy (GRE).

As metallic pipe is by far the most commonly used material used in a piping system, I concentrate on this sector. Circular in shape, pipe is identified in the various industry codes, standards, and specifications as a nominal pipe size (NPS), in U.S. customary units, or in *diamètre nominal* (DN) metric units, with a wall thickness quoted in one of the following ways:

- Standard weight (STD), extra strong (XS), double extra schedule (XXS).

- Carbon steel pipe in schedules, Sch 20, 30, 40, 60, 80, 120, 160.

- Stainless steel pipe in schedules, Sch 5S, 10S, 40S, 80S, 160S.

- Calculated wall thickness in U.S. customary units (inches) or metric units (mm), see Appendix B, Figure B–8 for a conversion chart.

Steel pipe is generally made by one of the following methods: seamless, longitudinally welded, or spirally welded. The first two are the most commonly used with seamless pipe available up to 24"; and longitudinally welded pipe generally is specified for sizes above 16", but it can be manufactured in smaller sizes.

Seamless pipe is formed by passing a solid billet with a mandrel through a metal bar that is at an elevated temperature. The bar is held between sizing rollers that dictate the outside diameter (O.D.) of the pipe, and the size of the billet creates the inside diameter (I.D.). Seamless pipe has a quality factor E of 1.0 in ASME B31.3 Table A-1B, presented here as Table 2–1.

Longitudinally welded pipe is created by feeding hot steel plate through shapers that roll the plate into a hollow circular section. The two edges of the pipe are squeezed together and welded. Longitudinally welded pipe has a quality factor E of 0.85 in ASME B31.3 Table A-1B.

Initially, longitudinal pipe has a lower integrity than seamless pipe; however, if the longitudinal weld is radiographically x-rayed successfully, then it is considered equal to seamless pipe with a quality factor E of 0.95, in ASME B31.3, Table A-1B.

Table 2-1 Basic Quality Factors for Longitudinal Weld Joints in Pipes, Tubes, and Fittings E_j

Spec. No.[a]	Class (or Type)	Description	E_j (2)	Appendix A Notes
Carbon Steel				...
API 5L	...	Seamless pipe	1.00	...
	...	Electric resistance welded pipe	0.85	...
	...	Electric fusion welded pipe, double butt, straight or spiral seam	0.95	...
	...	Furnace butt welded	0.60	...
A 53	Type S	Seamless pipe	1.00	...
	Type E	Electric resistance welded pipe	0.85	...
	Type F	Furnace butt welded pipe	0.60	...
A 105	...	Forgings and fittings	1.00	(9)
A 106	...	Seamless pipe	1.00	...
A 134	...	Electric fusion welded pipe, single butt, straight or spiral seam	0.80	...
A 135	...	Electric resistance welded pipe	0.85	...
A 139	...	Electric fusion welded pipe, straight or spiral seam	0.80	...
A 179	...	Seamless tube	1.00	...
A 181	...	Forgings and fittings	1.00	(9)
A 234	...	Seamless and welded fittings	1.00	(16)
A 333	...	Seamless pipe	1.00	...
	...	Electric resistance welded pipe	0.85	...
A 334	...	Seamless tube	1.00	...
A 350	...	Forgings and fittings	1.00	(9)
A 369	...	Seamless pipe	1.00	...

Table 2-1 Basic Quality Factors for Longitudinal Weld Joints in Pipes, Tubes, and Fittings E_j (cont'd)

Spec. No.[a]	Class (or Type)	Description	E_j (2)	Appendix A Notes
A 381	...	Electric fusion welded pipe, 100% radiographed	1.00	(18)
	...	Electric fusion welded pipe, spot radiographed	0.90	(19)
	...	Electric fusion welded pipe, as manufactured	0.85	...
A 420	...	Welded fittings, 100% radiographed	1.00	(16)
A 524	...	Seamless pipe	1.00	...
A 587	...	Electric resistance welded pipe	0.85	...
A 671	12, 22, 32, 42, 52	Electric fusion welded pipe, 100% radiographed	1.00	...
	13, 23, 33, 43, 53	Electric fusion welded pipe, double butt seam	0.85	...
A 672	12, 22, 32, 42, 52	Electric fusion welded pipe, 100% radiographed	1.00	...
	13, 23, 33, 43, 53	Electric fusion welded pipe, double butt seam	0.85	...
A 691	12, 22, 32, 42, 52	Electric fusion welded pipe, 100% radiographed	1.00	...
	13, 23, 33, 43, 53	Electric fusion welded pipe, double butt seam	0.85	...
Low and Intermediate Alloy Steel				
A 182	...	Forgings and fittings	1.00	(9)
A 234	...	Seamless and welded fittings	1.00	(16)
A 333	...	Seamless pipe	1.00	...
	...	Electric resistance welded pipe	0.85	...
A 334	...	Seamless tube	1.00	...

Table 2-1 Basic Quality Factors for Longitudinal Weld Joints in Pipes, Tubes, and Fittings E_j (cont'd)

Spec. No.[a]	Class (or Type)	Description	E_j (2)	Appendix A Notes
A 335	...	Seamless pipe	1.00	...
A 350	...	Forgings and fittings	1.00	...
A 369	...	Seamless pipe	1.00	...
A 420	...	Welded fittings, 100% radiographed	1.00	(16)
A 671	12, 22, 32, 42, 52	Electric fusion welded pipe, 100% radiographed	1.00	...
	13, 23, 33, 43, 53	Electric fusion welded pipe, double butt seam	0.85	...
A 672	12, 22, 32, 42, 52	Electric fusion welded pipe, 100% radiographed	1.00	...
	13, 23, 33, 43, 53	Electric fusion welded pipe, double butt seam	0.85	...
A 691	12, 22, 32, 42, 52	Electric fusion welded pipe, 100% radiographed	1.00	...
	13, 23, 33, 43, 53	Electric fusion welded pipe, double butt seam	0.85	...
Stainless Steel				
A 182	...	Forgings and fittings	1.00	...
A 268	...	Seamless tube	1.00	...
	...	Electric fusion welded tube, double butt seam	0.85	...
	...	Electric fusion welded tube, single butt seam	0.80	...
A 269	...	Seamless tube	1.00	...
	...	Electric fusion welded tube, double butt seam	0.85	...

Table 2–1 Basic Quality Factors for Longitudinal Weld Joints in Pipes, Tubes, and Fittings E_j (cont'd)

Spec. No.[a]	Class (or Type)	Description	E_j (2)	Appendix A Notes
	...	Electric fusion welded tube, single butt seam	0.80	...
A 312	...	Seamless tube	1.00	...
	...	Electric fusion welded tube, double butt seam	0.85	...
	...	Electric fusion welded tube, single butt seam	0.80	...
A 358	1, 3, 4	Electric fusion welded pipe, 100% radiographed	1.00	...
	5	Electric fusion welded pipe, spot radiographed	0.90	...
	2	Electric fusion welded pipe, double butt seam	0.85	...
A 376	...	Seamless pipe	1.00	...
A 403	...	Seamless fittings	1.00	...
	...	Welded fitting, 100% radiographed	1.00	(16)
	...	Welded fitting, double butt seam	0.85	...
	...	Welded fitting, single butt seam	0.80	...
A 409	...	Electric fusion welded pipe, double butt seam	0.85	...
	...	Electric fusion welded pipe, single butt seam	0.80	...
A 487	...	Steel castings	0.80	(9)(40)
A 789	...	Seamless tube	1.00	...
	...	Electric fusion welded, 100% radiographed	1.00	...

Table 2-1 Basic Quality Factors for Longitudinal Weld Joints in Pipes, Tubes, and Fittings E_j (cont'd)

Spec. No.[a]	Class (or Type)	Description	E_j (2)	Appendix A Notes
...	...	Electric fusion welded, double butt	0.85	...
...	...	Electric fusion welded, double butt	0.80	...
A 790	...	Seamless pipe	1.00	...
...	...	Electric fusion welded, 100% radiographed	1.00	...
...	...	Electric fusion welded, double butt	0.85	...
...	...	Electric fusion welded, single butt	0.80	...
A 815	...	Seamless fittings	1.00	...
...	...	Welded fittings, 100% radiographed	1.00	(16)
...	...	Welded fittings, double butt seam	0.85	...
...	...	Welded fittings, single butt seam	0.80	...
Copper and Copper Alloy				
B 42	...	Seamless pipe	1.00	...
B 43	...	Seamless pipe	1.00	...
B 68	...	Seamless tube	1.00	...
B 75	...	Seamless tube	1.00	...
B 88	...	Seamless water tube	1.00	...
B 280	...	Seamless tube	1.00	...
B 466	...	Seamless pipe and tube	1.00	...

Table 2–1 Basic Quality Factors for Longitudinal Weld Joints in Pipes, Tubes, and Fittings E_j (cont'd)

Spec. No.[a]	Class (or Type)	Description	E_j (2)	Appendix A Notes
B 467	...	Welded fittings, 100% radiographed	0.85	...
	...	Welded fittings, double butt seam	0.85	...
	...	Welded fittings, single butt seam	0.80	...
Nickel and Nickle Alloy				
B 160	...	Forgings and fittings	1.00	(9)
B 161	...	Seamless pipe and tube	1.00	...
B 164	...	Forgings and fittings	1.00	(9)
B 165	...	Seamless pipe and tube	1.00	...
B 167	...	Seamless pipe and tube	1.00	...
B 366	...	Seamless and welded fittings	1.00	(16)
B 407	...	Seamless pipe and tube	1.00	...
B 444	...	Seamless pipe and tube	1.00	...
B 464	...	Welded pipe	0.80	...
B 514	...	Welded pipe	0.80	...
B 517	...	Welded pipe	0.80	...
B 564	...	Nickel alloy forgings	1.00	(9)
B 619	...	Electric resistance welded pipe	0.85	...
	...	Electric fusion welded pipe, double butt seam	0.85	...
	...	Electric fusion welded pipe, single butt seam	0.80	...
B 622	...	Seamless pipe and tube	1.00	...
B 675	All	Welded pipe	0.80	...

Table 2-1 Basic Quality Factors for Longitudinal Weld Joints in Pipes, Tubes, and Fittings E_j (cont'd)

Spec. No.[a]	Class (or Type)	Description	E_j (2)	Appendix A Notes
B 690	...		1.00	...
B 705	...	Welded pipe	0.80	...
B 725	...	Electric fusion welded pipe, double butt seam	0.85	...
	...	Electric fusion welded pipe, single butt seam	0.80	...
B 729	...	Seamless pipe and tube	1.00	...
B 804	1, 3, 5	Welded pipe, 100% radiographed	1.00	...
	2, 4	Welded pipe, double fusion welded	0.85	...
	6	Welded pipe, single fusion welded	0.80	...
Titanium and Titanium Alloy				
B 337	...	Seamless pipe	1.00	...
	...	Electric fusion welded pipe, double butt seam	0.85	...
Zirconium and Zirconium Alloy				
B 523	...	Seamless tube	1.000	...
	...	Electric fusion welded pipe	0.80	
B 658	...	Seamless pipe	1.00	...
	...	Electric fusion welded pipe	0.80	...
Aluminum Alloy				
B 210	...	Seamless tube	1.00	...
B 241	...	Seamless pipe and tube	1.00	...
B 247	...	Forgings and fittings	1.00	(9)

Table 2–1 Basic Quality Factors for Longitudinal Weld Joints in Pipes, Tubes, and Fittings E_j (cont'd)

Spec. No.[a]	Class (or Type)	Description	E_j (2)	Appendix A Notes
B 345	...	Seamless pipe and tube	1.00	...
B 361	...	Seamless fittings	1.00	...
	...	Welded fittings, 100% radiograph	1.00	(18)(23)
	...	Welded fittings, double butt	0.85	(23)
	...	Welded fittings, single butt	0.80	(23)
B 547	...	Welded pipe and tube, 100% radiograph	1.00	...
	...	Welded pipe, double butt seam	0.85	...
	...	Welded pipe, single butt seam	0.80	...

a. Printed with the permission of the ASME.

Spiral welding is the least common method of manufacturing pipe. It is formed by twisting strips of metal into a spiral pattern. This type of pipe is the cheapest, and it generally is used only for piping systems in nontoxic service, such as cooling water at atmospheric or very low pressures and for very large sizes. Spirally welded pipe has a quality factor *E* of 1.0 in ASME B31.3 Table A-1B.

The quality factor *E* is used in the formula in ASME B31 codes to calculate the wall thickness of pressure containing pipe. So that a higher *E* factor in the calculation results in a thinner and therefore lighter pipe.

This formula can be found in ASME B31.3 (304, Pressure Design of Components; 304.1, Straight Pipe; 304.1.1, General):

$$t = \frac{PD}{2(SE + PY)} \qquad (2.1)$$

where

> P is the internal design gauge pressure.
>
> D is the outside diameter of pipe as listed in tables of standards or specifications or as measured.
>
> S is the stress value for material from Table A-1.
>
> E is the quality factor from Table A-1B.
>
> P is the internal design gauge pressure.
>
> Y is the coefficient from Table 304.1.1, valid for $t\,D/6$ and for the materials shown.

The value of Y may be interpolated for intermediate temperatures. For $t\,D/6$,

$$Y = \frac{d+2c}{D+d+2c} \qquad (2.2)$$

The lower the quality factor E, the greater the wall thickness will be calculated in these formulae. This increases the amount of material required for the pipe and increases its weight. Radiography comes at a price, but it raises the quality factor E to 1.00, which results in thinner wall thickness and a reduction in weight. The general option is to radiograph the longitudinal pipe.

All three methods have advantages and disadvantages, both commercially and technically. Longitudinal pipe can be manufactured to closer tolerances than seamless pipe, but it requires additional radiography to bring it up to the same quality factor as seamless pipe.

Generally, steel pipe is produced in two lengths: single random (SRL) and double random (DRL). Taken from API 5L, Specification for Line Pipe, nominal lengths of 20 ft (6 m) formerly were designated single random lengths and those of 40 ft (12 m) double random lengths.

If long runs are anticipated, as on a pipe rack, DRL is preferred, because it will result in fewer field welds. If this is not an issue, the shorter SRL are an option. NPS 1½" and smaller pipe sizes are too whippy at DRL and are supplied in SRL.

2.2.1 Pipe Sizes

During the early years of the oil and gas industry in the United States, the dimensional system was known as *iron pipe size* (IPS). The size identified the approximate inside diameter of the pipe in inches. For example IPS 6 pipe has an inside diameter of approximately 6 inches. These IPS sizes had the wall thicknesses that were identified as

> Standard weight (STD WT) for lower pressure piping—ASME class 150 and 300.
>
> Extra strong (XS) or extra heavy (XH) for medium pressure—ASME class 600.
>
> Double extra strong (XXS) or double extra heavy (XXH) for high pressure—ASME class 900 and above.

As the oil and gas industry developed and more sophisticated new materials were developed and became available, such as carbon steel in its various guises, low and intermediate alloys, and corrosion-resistant alloys (CRA) like stainless steel, the original dimensional system required updating to accommodate the improved characteristics that these new materials brought.

Corrosion-resistant alloys meant that corrosion allowances could be reduced and, in many cases, dispensed with, resulting in a reduction of the wall thicknesses and less weight. These thinner walls required a new method to identify the size and the wall thickness of this expanded range of pipes. The new designation, known as *"nominal pipe size,"* replaced the terminology IPS, and the term *schedule* (SCH) was applied to specify the nominal wall thickness of pipe.

Essentially, nominal pipe size is a dimensionless designator of pipe size. It identifies a pipe size without an inch symbol. For example, NPS 2 indicates a pipe whose outside diameter is 2.375 in. The NPS 12 and smaller pipes have outside diameters greater than the size designator (2, 4, 6, 8, 10, 12). However, the outside diameter of NPS 14 and larger pipes is the same as the size designator in inches. For example, NPS 14 pipe has an outside diameter equal to 14 in. The inside diameter depends on the pipe-wall thickness specified by the schedule number. Refer to ASME B36.10M or ASME B36.19M.

In spite of the introduction of the "schedule" method of identifying the wall thickness of a pipe, STD, XS(XH), and XXS (XXH) still are commonly used within the industry, and at certain sizes, there is a correlation between STD/Sch40, XS/Sch80, and XXS/Sch160, but always check and never guess.

Diamètre nominal is the dimensionless designator of pipe size in the metric unit system, developed by the International Standards Organization (ISO).

Table 2–2 is a cross reference between NPS U.S. customary units (inches) and DN metric units (millimeters).

Dimensional Specifications

Pipes used within a plant designed to one of the various ASME B31 codes generally are manufactured to a set of requirements specified in one of two American Society for Mechanical Engineers standards, depending on the materials of construction:

> B36.10M, Welded and Seamless Wrought Steel Pipe: Carbon Steel, Low Temperature Carbon Steel.
>
> B36.19M, Stainless Steel Pipe: Stainless Steel and High Alloy.

Standard B36.10M, Welded and Seamless Wrought Steel Pipe, covers the standardization of dimensions of welded and seamless wrought steel pipe for high or low temperatures and pressures. The word *pipe*, as distinguished from *tub*e, is used to apply to tubular products of dimensions commonly used for pipeline and piping systems. Pipe NPS 12 (DN 300) and smaller have outside diameters numerically larger than corresponding sizes.

Standard B36.19M, Stainless Steel Pipe, covers the standardization of dimensions of welded and seamless wrought stainless steel pipe. The word *pipe*, as distinguished from *tube*, is used to apply to tubular products of dimensions commonly used for pipeline and piping systems. Pipe dimensions of sizes 12 and smaller have outside diameters numerically larger than the corresponding size.

Another piping dimensional specification is American Petroleum Institute's API 5L, which is for carbon steel line pipe, and this document

Table 2–2 *Diamètre Nominal* (DN) and Nominal Pipe Size (NPS)

Diamètre Nominal, DN (mm)[a]	Nominal Pipe Size, NPS (inches)
6	
8	¼
10	
15	½
20	¾
25	1
32	1¼
40	1½
50	2
65	2½
80	3
100	4
150	6
200	8
250	10
300	12
350	14
400	16
450	18
500	20
550	22
600	24
650	26
700	28
750	30

Table 2–2   *Diamètre Nominal* (DN) and Nominal Pipe Size (NPS) (cont'd)

Diamètre Nominal, DN (mm)[a]	Nominal Pipe Size, NPS (inches)
800	32
900	36
1000	40
1100	42
1200	48
1400	54
1500	60
1600	64
1800	72
2000	80
2200	88

a. The size of pipes, fittings, flanges and valves are given in either millimeters as DN (metric units) or inches as NPS (U.S. customary units).

tracks ASME B36.10/19 dimensionally but goes into more detail covering additional subjects, such as material requirements, inspection, and testing:

1. Scope.
2. References.
3. Definitions.
4. Information to be Supplied by the Purchaser.
5. Process of Manufacture and Material.
6. Material Requirements.
7. Dimensions, Weights, Lengths, Defects, and End Finishes.
8. Couplings (PSL 1 Only).

9. Inspection and Testing.

10. Marking.

11. Coating and Protection.

12. Documents.

13. Pipe Loading.

Material Specifications

The most commonly used pipe material in the oil and gas industry is carbon steel (CS) and a chemically modified version for operating at temperatures down to –46°C, aptly called *low-temperature carbon steel* (LTCS). Both versions of carbon steel combine strength and a basic level of resistance to corrosive services. In slightly more corrosive service, an additional calculated allowance can be added to the wall thickness of the pipe, called a *corrosion allowance* (CA). The corrosion allowance increments usually are $^1/_{16}$" (1.5 mm), $^1/_8$" (3 mm), or ¼" (6 mm) and it is very rare for the CA to exceed ¼", because a more corrosion-resistant metal will be specified.

The Corrosion Allowance Increments

Process piping also can be supplied in a variety of corrosion-resistant alloys, such as stainless steel or nickel alloy and various other materials with specialized chemical compositions. These lesser-used nickel alloys are called *exotic* materials because of their rarity; they are used for very special services that have particularly corrosive characteristics at both ambient and elevated temperatures.

A list of piping component materials is referenced in ASME B31.3 and are detailed in Appendix A, under the very obvious industry-recognized title, "Listed Material." This list identifies materials that can be used on a project designed to the code, without the necessity to verify the mechanical or the physical properties. Material not on this list, known as *unlisted material*, can be used, but there is a requirement to authenticate the material data sheet.

Pipe can be manufactured using various processes; however, the most commonly specified that satisfy the requirements of ASME B31.3 are seamless, welded longitudinally (EFW, SAW, DSAW), and welded spirally. The method of manufacturing depends on size, process requirements, and economics.

2.2.2 Pipe Ends

Pipe ends can be supplied in several variations; these are the most commonly specified within ASME B31.3:

- Plain end (PE), usually pipe 2" and below.

- Threaded end (TE), usually pipe 2" and below.

- Butt weld (BW) or weld end (WE), all sizes.

Dimensional Standards for Pipe Ends

Plain end pipe is simply a cut 90° perpendicular to the outside diameter of the pipe that passes through the centerline of the pipe to the opposite side. It is also called a *square cut*, because of the 90° angle. Plain end pipe can be reprepared, also called *reprepped* for short, to form either threaded or butt-weld ends.

There is no standard for plain end pipe, because of its simple geometry: however, threaded end and butt-weld ends are more complex and the geometry, dimensions, and tolerances of these two examples are covered in the following ASME standards.

Threaded Ends

A threaded end joint also has a specific geometry, depending on the wall thickness of the pipe; and this is specified in ASME B1.20.1 (see Appendix B, Figure B–11). To connect to lengths of straight pipe, a coupling with matching threads is required. The dimensional design of a full coupling is covered by ASME B16.11, but the internal threads are governed by ASME B1.20.1, 1983, Pipe Threads, General Purpose, Inch. This American National Standard covers dimensions and gauging of pipe threads for general purpose applications. The B1.20.1 is a revision and redesignation of ANSI B2.1, 1968. The general purpose types are available to satisfy particular requirements, and manufacturers should be consulted for special requirements. The inclusion of dimensional data in this standard is not intended to imply that all of the products described are stock production sizes. Consumers are requested to consult with manufacturers concerning availability of products.

The standard covers the following areas:

1. Introduction.

    1.1 Scope.

    1.2 Thread Designations.

    1.3 Sealing.

    1.4 Inspection.

    1.5 Appendix.

    1.6 Related Standard.

2. American National Standard Pipe Thread Form.

    2.1 Thread Form.

    2.2 Angle of Thread.

    2.3 Truncation and Thread Height.

3. Specification for General Purpose Taper Pipe Threads, NPT.

    3.1 Taper Pipe Threads.

    3.2 Tolerances.

4. Specifications for Internal Straight Threads in Pipe Couplings, NPSC.

    4.1 Straight Pipe Threads in Pipe Couplings.

5. Specifications for Railing Joint Taper Pipe Threads, NPTR.

    5.1 Railing Joints.

6. Specifications for Straight Pipe Threads for Mechanical Joints: NPSM, NPSL, NPSH.

    6.1 Straight Pipe Threads.

    6.2 Free-Fitting Mechanical Joints for Fixtures, NPSM.

    6.3 Loose-Fitting Mechanical Joints With Locknuts, NPSL.

    6.4 Loose-Fitting Mechanical Joints for Hose Coupling, NPSH.

7. Gauges and Gauge Tolerances for American National Standard Pipe Threads.

    7.1 Design of Gauges.

    7.2 Classes of Gauges.

    7.3 Gauge Tolerances.

    7.4 Relation of Lead and Angle Deviations to Pitch Diameter, Tolerances of Gauges.

8. Gauging of Taper Pipe Threads.

    8.1 Gauging External Taper Threads.

    8.2 Gauging Internal Taper Threads.

    8.3 Gauging Practice.

    8.4 Gauging Chamfered, Countersunk, or Recessed Threads.

9. Gauging of Straight Pipe Threads.

    9.1 Types of Gauges.

    9.2 Gauge Dimensions.

Butt-Weld Ends

A butt-weld (or weld-end) joint also has specific geometry depending on the wall thickness of pipe; and this is specified in ASME B16.25, Butt Welding Ends, which covers the following subjects:

>Committee Roster.
>
>Correspondence with the B16 Committee.
>
>Scope.
>
>Transition Contours.
>
>Welding Bevel Design.
>
>Preparation of Inside Diameter of Welding End.
>
>Tolerances.
>
>Figures.
>
>Table.
>
>Mandatory Appendices.
>
>Nonmandatory Appendix.

This covers straight pipe; however, to facilitate the mechanical jointing, the changes of direction, changes in O.D. pipe size, and the merging two pipes, numerous other piping components are required to create and complete a process piping system.

2.3 Pipe Fittings

Pipe fitting components complement straight pipe, and within a piping system, both must be chemically and mechanically compatible. Pipe fitting components are used for one or more functions:

- Change of direction—90° and 45° elbows.
- Change of direction—equal tee.

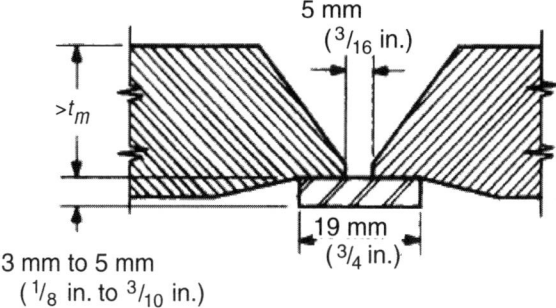

Butt Joint With Bored Pipe Ends and Solid or Split Backing Ring

Figure 2–2 *Bevels for wall thickness over 3 mm (0.12 in.) to 22 mm (0.88 in.), inclusive. (Printed with the permission of ASME.)*

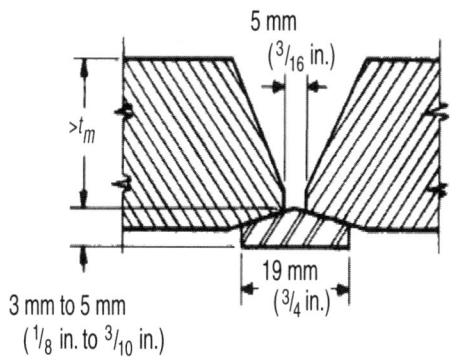

Butt Joint With Taper-Borred Ends and Solid Backing Ring

Figure 2–3 *Well bevel details for wall thickness over 22 mm (0.88 in.). (Printed with the permission of ASME.)*

- Reduction in pipe size—eccentric and concentric reducers, swages.
- Reduction in pipe size and change of direction—reducing tee.

- Pipe joint—flange, coupling, union.

- Reinforced branch fitting—Weldolet, Sockolet, Threadolet.

- Mechanical joints—flanges.

Pipe fittings used for projects designed to ASME B31 code are made to standard dimensions, based on their size and wall thickness. These fixed dimensions are essential to allow a piping designer to lay out or route the piping system efficiently.

All these piping components can be joined together by several welding and mechanical methods: butt-weld, socket weld or threaded ends, flanges (bolts and gaskets), or proprietary mechanical joints (Victaullic, hub ends).

2.3.1 Butt-Weld End Fittings

Butt-weld fittings have beveled ends (BE), specially prepared to ASME B16.25, that allow a high-integrity, full-penetration, circumferential butt-weld to be completed and join the fitting to pipe, fitting to fitting, fitting to valve, and valves to pipe. At the beginning of the welding process the two butt-weld ends are place together with a $1/16"$ root gap. This gap is occupied with the first pass or filler material. Nondestructive examination (NDE) can be carried out to guarantee that this weld is sound and defect free.

Butt-weld fittings include elbows, tees (full and reducing), and eccentric and concentric reducers covered in ASME B16.9. Fittings to this standard usually are supplied with butt-weld ends to ASME B16.25.

B16.9, Factory-Made Wrought Butt Welding Fittings, defines overall dimensions, tolerances, ratings, testing, and markings for wrought carbon and alloy steel factory-made butt-weld fittings of NPS ½" through 48" DN 15 to1200. It covers fittings of any producible wall thickness. The standard covers the following subjects:

1. Scope.

2. Pressure Ratings.

3. Size.

4. Marking.

5. Material.

6. Fitting Dimensions.

7. Surface Contours.

8. End Preparation.

9. Design Proof Test.

10. Production Tests.

11. Tolerances.

2.3.2 Socket-Weld and Threaded-End Fittings

Socket-weld and threaded-end fittings generally are used in smaller sizes (NPS 2", DN 50 and below), for utility low pressure (51.0 barg and below), and media (like water, nitrogen, air, and certain noncorrosive agents) service at ambient and slightly elevated temperatures. Socket welds are male/female connections created by a circumferential single-fillet weld, and they are not of the same high integrity as the butt-weld but an easier weld to perform and cheaper to fabricate. To give added confidence, this fillet weld can be subjected to NDE to raise the level of confidence in the weld.

A screwed connection is a male/female mechanical, nonwelded joint and, therefore, not a permanent method of joining, which has advantages if the piping system has to be disassembled. Threaded connections are limited to low pressures and should not be used at elevated temperatures, where bending moments are expected, or cyclic conditions, because the geometry of the threaded connection might become distorted.

Socket-weld and threaded fittings include elbows, tees (full and reducing), eccentric and concentric reducers, couplings, and the like that are covered in ASME B16.11, Forged Fittings Socket Welding and Threaded. Fittings to this standard usually are supplied with threaded ends to ASME B1.20.1. The B16.11 is a dimensional specification for forged fittings, socket weld and threaded, that covers ratings, dimensions, tolerances, marking, and material requirements for forged fittings, both socket weld and threaded, from NPS 1/8" to 4". This document covers the following subjects:

> Committee Roster.
>
> Correspondence with the B16 Committee.
>
> (1) Scope.
>
> (2) Pressure Ratings.
>
> (3) Size and Type.
>
> (4) Marking.
>
> (5) Material.
>
> (6) Dimensions.
>
> (7) Tolerances.
>
> (8) Testing Figures.

2.3.3 Flanged Joints

Flanged joints are a mechanical, nonpermanent method of joining two flanged piping components and one of the most commonly used methods of joining together pipe to pipe, pipe to fitting, and pipe to valve. It is a mechanical joint that, if assembled correctly, using the correct components and the right bolting procedure, results in a leak-free connection that can be dismantled and reassembled, if necessary.

A flange is an integral fitting with two distinct areas;

- The flange blade with the bolt holes and the sealing face.
- The flange hub with the pipe connection ends.

The flange blade is the circular area through which there is a standard bolting pattern, based on the O.D. size of the pipe and the design pressure rating. It has a seal face accurately machined to a predetermined finish, on which the gasket sits and a flange back on which the nut sits. The hub is located on the back of the blade and it receives the pipe.

Several methods are used to attach flanges to other fittings, each with technical and commercial merits:

- Weld neck flanges—attached by one butt-weld, high integrity.
- Socket weld flanges—attached by one socket weld, medium integrity.
- Threaded flanges—attached by one threaded end, low integrity.
- Slip-on flanges—attached by one or sometimes two fillet welds, medium integrity.
- Lap joint n (stub end) flanges—attached by one butt-weld on the stub end, high integrity.
- Blind flanges—attached by a mechanical bolt up to any mating flange.

Weld Neck Flanges

Weld neck (WN) flanges are available at all sizes and ratings, and they offer the best alternative for combined high integrity, medium installation cost, and standardization. They come in a variety of flange facings, including the three most commonly available: raised face (RF), with low-, medium-, and high-pressure classes; flat face (FF), with a low-pressure class; and ring-type joint (RTJ), with low-, medium-, high-, and very high-pressure classes.

The weld neck flange is an integral one-piece component, with two distinct parts: the hub and the blade. The hub has a tapered neck with one end prepared for a butt-weld connection to the pipe and the other end that reinforces and supports the "blade" of the flange and prevents dishing or bowing at elevated temperatures and pressure. The blade has a drilling pattern that allows it to be mated against other compatible flanges. The weld neck design and the high-integrity butt-weld make this the most robust option for a flange that will be subjected to elevated temperatures and pressures.

The butt-weld can be examined using magnetic particle inspection (MPI), dye penetrant inspection (DPI), radiography, or ultrasonic inspection.

The flange facing most commonly used for weld-neck flanges is the raised face. The gasket seating surface is a circular raised ¼" platform above the bolting circle face. The pressure class of the flange dictates the height of the raised face according to ASME B16.5. The flat-face or "full-face" flange has a gasket surface in the same plane as the bolting circle face. Applications using flat-face or full-face flanges frequently are those in which the mating flange or flanged fitting is made from a casting and the flush mating means no possibility for the flange blade to bow and crack or deform.

Raised and flat flange facings are machined, and they may be either phonographic (spiral) serrated or concentric serrated. *Phonographic* means that the finishing groove spirals in toward the center of the flange blade and a concentric grooving means a series of unconnected concentric grooves on the face of the flange. The industry norm is a phonographic serrated finish.

The facing finish is measured by visual comparison with roughness average (Ra) standards. Ra is stated in micro inches (μin) or micrometers (μm) and shown as an arithmetic average roughness height (AARH) or root mean square (rms). AARH and rms are different methods of calculation giving essentially the same result and are used interchangeably for these products.

The micro profile on the flange face bites into the soft gasket that is trapped between the other mating flanges by the compressive forces applied during the bolt-up.

The industry standard Ra supplied by manufacturers is 125 to 250 μin or 3.2 to 6.3 μm AARH or rms. The short form is 125–250 AARH or 3.2–6.4 AARH. Other finishes are available at the customer's request.

The gasket contact surface for a ring-type joint flange is inside the groove cut into the face. The steel ring gasket fits into the grooves of the mating flanges and is sealed with pressure. The finish in the ring grooves and on the ring gasket is 63 μin AARH or 1.6 μm AARH maximum.

Socket-Weld Flanges

Socket-weld flanges are available up to 4" NPS, but the most commonly used size range is ½–2" NPS (Figure 2–4). The pipe is inserted into the socket hub and fillet welded into place. Care must be taken

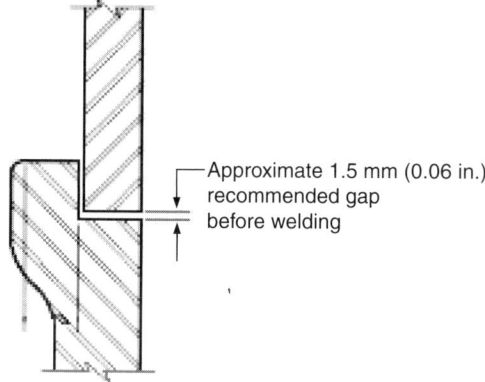

Figure 2-4 *Recommended gap for socket weld fit up, prior to welding.*

not to push the pipe too far into the socket of the hub so that it "bottoms" and, during the welding process, the hot pipe expands and deforms because it stopped by the base of the socket.

Radiography is not practical on the fillet weld; therefore, correct fitting and welding is crucial. The fillet weld may be inspected by surface examination, magnetic particle, or liquid penetrant examination methods.

The fillet weld used to attach the pipe to the flange is not considered a high-integrity weld, and NDE is not so easy to perform. Hence, the use of socket weld flanges is restricted to low- and medium-pressure classes, up to ASME 600 class. The flange facings also usually are restricted to raised-face and flat-faced flanges.

Threaded Flanges

Threaded flanges are generally used in the size range ½–2" and usually only for utility services such as air, water, or nitrogen at low pressures, up to ASME 300 class. Use of these flanges at elevated temperatures is not recommended, because the geometry of the thread may deform at elevated temperatures. Because it is a screwed connection, it lacks the integrity of either a butt-weld or a socket-weld joint. An advantage is that the threaded connection is not permanent and it can be disassembled. The integrity of this connection can be improved by seal welding using fillet weld; however, this makes it a permanent joint.

Like the socket-weld flange, the flange facings usually are limited to raised-face (low-, medium-, and high-pressure classes) and flat-face (low- and medium-pressure classes) flanges.

Lap-Joint Flanges with a Stub End

A lap-joint flange is a two-component assembly, with a stub end that has a lap-joint ring flange placed over it. The stub end is then butt welded to the pipe, and the flange ring can be rotated to align with the mating flange. This type of flange connection is particularly useful for large or hard-to-adjust flanges. The lap joint flange can be used in sizes and pressure classes similar to that of a weld-neck flange.

The nature of this joint means that the stub end facing is also the flange facing, which makes it raised faced, and the gasket seating surface.

Like the weld-neck fitting, the lap-joint flange butt-weld connection can be examined using magnetic particle inspection, dye penetrant inspection, radiography, or ultrasonic inspection.

Slip-on Flanges

The slip-on flange has a very low-profile hub, through which the pipe is passed. Generally, two fillet welds are performed, one internal and one external. Although the initial cost of a slip-on flange is less than a weld neck, by the time the two fillet welds have been performed, there is very little difference in the cost. Also the slip-on flange with two fillet welds requires two NDEs, while the weld neck requires one NDE.

Generally, the slip-on flange is available in similar sizes as a weld-neck flanges, but it is not commonly used above ASME class 600.

Blind Flange

A blind flange is a closure plate flange that terminates the end of a piping system. It can be used in combination with all of the previous flanges at all sizes and all pressure classes. It comes in the following facings: raised faced (low-, medium-, and high-pressure classes), flat faced (low-pressure class), and ring-type joint (low-, medium-, high-, and very high-pressure classes).

Dimensional Standards for ASME Flanges

The dimensional standards for the various types of flanges just mentioned are covered in two ASME standards: ASME B16.5, NPS ½" through 24" (DN 15 through 600) and ASME B16.47, NPS 26" through 60" (DN 650 through 1500). Depending on the jointing method between the flange and the pipe, one of the following ASME dimensional standards will apply: ASME B16.25 for butt-weld ends, ASME B1.20.1 for threaded ends, or ASME B16.11 for socket-weld ends.

Standard B16.5, 2003, Pipe Flanges and Flanged Fittings: NPS ½" through 24", covers pressure-temperature ratings, materials, dimensions, tolerances, marking, testing, and methods of designating openings for pipe flanges and flanged fittings. Included are flanges with rating class designations 150, 300, 400, 600, 900, 1500, and 2500 in sizes NPS ½ through NPS 24, with requirements given in both metric and U.S. customary units with diameter of bolts and flange bolt holes expressed in inch units; flanged fittings with rating class designation 150 and 300 in sizes NPS ½ through NPS 24, with requirements given in both metric and U.S. customary units, with diameter of bolts and flange bolt holes expressed in inch units; and flanged fittings with rating class designation 400, 600, 900, 1500, and 2500 in sizes NPS ½ through NPS 24 that are acknowledged in Annex G, in which only U.S. customary units are provided.

This standard is limited to flanges and flanged fittings made from cast or forged materials, blind flanges, and certain reducing flanges made from cast, forged, or plate materials.

Also included in this standard are requirements and recommendations regarding flange bolting, flange gaskets, and flange joints. The subject matter is as follows:

> Committee Roster.
>
> Correspondence with the B16 Committee.
>
> (1) Scope.
>
> (2) Pressure-Temperature Ratings.
>
> (3) Component Size.
>
> (4) Marking.

(5) Materials.

(6) Dimensions.

(7) Tolerances.

(8) Pressure Testing.

Standard B16.47, Large Diameter Steel Flanges, covers pressure-temperature ratings, materials, dimensions, tolerances, marking, and testing for pipe flanges in sizes NPS 26 through NPS 60 and in ratings classes 75, 150, 300, 400, 600, and 900.

Flanges may be cast, forged, or plate (for blind flanges only) materials, as listed in Table 1A. Requirements and recommendations regarding bolting and gaskets are included. The subject matter is as follows:

Standards Committee Roster.

(1) Scope.

(2) Pressure-Temperature Ratings.

(3) Size.

(4) Marking.

(5) Materials.

(6) Dimensions.

(7) Tolerances.

(8) Test.

2.4 Valves

Valves are the most complex component within a piping system and I cover the fundamentals of their design and construction in this chapter. Unlike pipe and piping fittings, valves are multicomponent items, with a variety of materials of construction and static (stationery) and dynamic (moving) parts. They are a vital part of a piping system and, depending on their design, are capable of transporting liquids, gases, vapors, and slurries.

The value of valves, both commercially and functionally, is greatly overlooked in a process plant. Valves are the controlling element of process flow: They start, stop, regulate, check, and come in a variety of materials of construction and design types.

The origins of valves can be traced back to the Romans, who used what would be called a *plug-type valve* to start, stop, and divert the flow of water in channels and pipes.

The most commonly used valves in projects designed to the ASME B31 code are

- Gate valves.
- Globe valves.
- Check valves.
- Ball valves.
- Plug valves.
- Butterfly valves.
- Pinch or diaphragm valves.
- Control valves.
- Pressure relief valves.
- Control valves.

Each of these can be subdivided in other groupings based on their design and materials of construction.

Valves can be operated either manually, by operating personnel, or using an independent power source, either electric, pneumatic, or hydraulic, depending on the power requirement and availability.

A valve is a multicomponent item that has both dynamic (moving) and static (nonmoving) parts. It can be constructed in metallic or nonmetallic materials. A valve can perform one or more of the following functions:

- Start/stop flow (butterfly valve)—isolating valve, such as a gate, ball, or plug valve.
- Regulate flow (butterfly valve)—throttle or globe valve.
- Prevent backflow—nonreturn or check valve.
- Control flow—control valve.

Valves selected for ASME B31 code projects are governed by numerous international standards and specifications, which have been created to ensure that the valve selected will function predictably and the possibility of in service malfunction is avoided.

These standards cover the type of valve, design, construction, components, dimensions, testing, and marking.

2.4.1 Valve Codes and Standards

The material that follows includes the most commonly used standards for valves used in ASME B31 process projects from a variety of recognized bodies. These codes and standards contain the rules and requirements for design, pressure-temperature ratings, dimensions, tolerances, materials, nondestructive examinations, testing, and inspection and quality assurance. Compliance to these and other standards is invoked by reference to codes of construction, specifications, contracts, or regulations.

ASME Standards

B16.10, Face-to-Face and End-to-End Dimensions of Valves.

B16.20, Metallic Gaskets for Pipe Flanges: Ring Joint, Spiral Wound, Jacketed.

B16.21, Non Metallic Flat Gaskets for Pipe Flanges.

B16.34, Valves—Flanged, Threaded, and Welding End.

B16.38, Large Metallic Valves for Gas Distribution (Manually Operated, NPS 2½" to 12, 125 psig Maximum).

B16.40, Manually Operated Thermoplastic Gas Shutoffs and Valves in Gas Distribution Systems.

AWWA Standards and Specifications

C500, Metal-Seated Gate Valves for Water Supply Service.

C501, Cast-Iron Sluice Gates.

C504, Rubber-Seated Butterfly Valves.

C507, Ball Valves, 6 in. through 48 in. (150 mm through 1200 mm).

C508, Swing Check Valves for Waterworks Service.

C509, Resilient-Seated Gate Valves for Water Supply Service.

C510, Double Check Valve Backflow Prevention Assembly.

C511, Reduced Pressure Principle Backflow Prevention Assembly.

C512, Air-Release, Air-Vacuum, and Combination Air Valves for Waterworks Service.

C540, Power-Actuating Devices for Valves and Sluice Gates.

C550, Protective Epoxy Interior Coatings for Valves and Hydrants.

American Petroleum Institute Specifications

6D, Specification for Pipeline Valves (Gate, Plug, Ball, and Check Valves).

6FA, Specification for Fire Test for Valves.

6FB, Specification for Fire Test for End Connections.

6FC, Specification for Fire Test for Valves with Automatic Backseats.

6FD, Specification for Fire Test for Check Valves.

14A, Specification for Subsurface Safety Valve Equipment.

14D, Specification for Wellhead Surface Safety Valves and Underwater Safety Valve for Offshore Service.

API Standards

526, Flanged Steel Pressure Relief Valves.

527, Seat Tightness of Pressure Relief Valves.

589, Fire Test for Evaluation of Valve Stem Packing.

594, Wafer and Wafer-Lug Check Valves.

598, Valve Inspection and Testing.

599, Metal Plug Valves—Flanged and Welding Ends.

600, Steel Gate Valves—Flanged and Butt-Welding Ends.

602, Compact Steel Gate Valves—Flanged, Threaded, Welding, and Extended-Body Ends.

603, Class 150, Cast, Corrosion-Resistant, Flanged-End Gate Valves.

607, Fire Test for Soft-Seated Quarter-Turn Valves.

608, Metal Ball Valves—Flanged, Threaded, and Welding Ends.

609, Lug- and Wafer-Type Butterfly Valves.

MSS Standards

MSS-SP-6, Standard Finishes for Contact Faces of Pipe Flanges and Connecting-End Flanges of Valves and Fittings.

MSS-SP-25, Standard Marking System for Valves, Flanges, and Fittings.

MSS-SP-42, Class 150 Corrosion Resistant Gate, Globe, Angle, and Check Valves with Flanged and Butt-Weld Ends.

MSS-SP-45, Bypass and Drain Connection Standard.

MSS-SP-53, Quality Standard for Steel Castings and Forgings for Valves, Flanges, and Fittings and Other Piping Components—Magnetic Particle Examination Method.

MSS-SP-54, Quality Standard for Steel Castings and Forgings for Valves, Flanges, and Fittings and Other Piping Components—Radiographic Examination Method.

MSS-SP-55, Quality Standard for Steel Castings and Forgings for Valves, Flanges, and Fittings and Other Piping Components—Visual Method.

MSS-SP-60, Connecting Flange Joint between Tapping Sleeves and Tapping Valves.

MSS-SP-61, Pressure Testing of Steel Valves.

MSS-SP-67, Butterfly Valves.

MSS-SP-68, High Pressure-Offset Seat Butterfly Valves.

MSS-SP-70, Cast Iron Gate Valves, Flanged and Threaded Ends.

MSS-SP-71, Cast Iron Swing Check Valves, Flanged and Threaded Ends.

MSS-SP-72, Ball Valves with Flanged or Butt-Welding Ends for General Service.

MSS-SP-78, Cast Iron Plug Valves, Flanged and Threaded Ends.

MSS-SP-80, Bronze Gate, Globe, Angle and Check Valves.

MSS-SP-81, Stainless Steel, Bonnetless, Flanged Knife Gate Valves.

MSS-SP-82, Valve Pressure Testing Methods.

MSS-SP-84, Valves—Socket-Welding and Threaded Ends.

MSS-SP-85, Cast Iron Globe and Angle Valves, Flanged and Threaded Ends.

MSS-SP-86, Guidelines for Metric Data in Standards for Valves, Flanges, Fittings, and Actuators.

MSS-SP-88, Diaphragm Type Valves.

MSS-SP-91, Guidelines for Manual Operation of Valves.

MSS-SP-92, MSS Valve User Guide.

MSS-SP-93, Quality Standard for Steel Castings and Forgings for Valves, Flanges, and Fittings and Other Piping Components—Liquid Penetrant Examination Method.

MSS-SP-94, Quality Standard for Steel Castings and Forgings for Valves, Flanges, and Fittings and Other Piping Components—Ultrasonic Examination Method.

MSS-SP-96, Guidelines on Terminology for Valves and Fittings.

MSS-SP-98, Protective Epoxy Coatings for the Interior of Valves and Hydrants.

MSS-SP-99, Instrument Valves.

MSS-SP-100, Qualification Requirements for Elastomer Diaphragms for Nuclear Service Diaphragm Type Valves.

MSS-SP-101, Part–Turn Valve Actuator Attachment—Flange and Driving Components Dimensions and Performance Characteristics.

MSS-SP-102, Multi-Turn Valve Actuator Attachment—Flange and Driving Component Dimensions and Performance Characteristics.

MSS-SP-105, Instrument Valves for Code Applications.

MSS-SP-108, Resilient Seated-Eccentric Cast Iron Plug Valves.

2.4.2 Classification of Operation Valves

Valves have three basic methods of functional operation, based on the movement of their closure member: linear, rotary, or automatic. A valve whose closure member moves in a straight line to the open or closed position is linear. This includes gate, globe, and diaphragm valves. A valve whose closure member travels rotationally from the fully open to the fully closed position, usually in 90°, is rotary; it is also known as a *quarter-turn valve*. A valve whose closure member moves without manual or motorized assistance is automatic. This includes check valves, such as piston lift, swing, dual plate, and relief valves. Table 2–3 summarizes the types of valves.

Table 2–3 Types of Operation Valves

	Linear	Rotary	Automatic
Gate valve	See Figure 2–5		
Globe valve	See Figure 2–6		
Swing check valve			See Figure 2–7
Lift check valve			See Figure 2–8
Tilting-disc check valve			See Figure 2–9
Dual-plate check valve			See Figure 2–10
Ball valve		See Figure 2–11	
Pinch valve	See Figure 2–12		
Butterfly valve		See Figure 2–13	
Plug valve		See Figure 2–14	
Diaphragm valve	See Figure 2–15		

88 Chapter 2—Piping Components

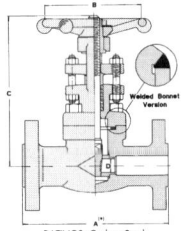

RATINGS: Carbon Steel
Class 150 - 285 p.s.i. @ 100°F
Class 300 - 740 p.s.i. @ 100°F
Class 600 - 1480 p.s.i. @ 100°F

CLASS 150-300-600
BOLTED BONNET - REGULAR PORT - ISO 15761
Outside Screw & Yoke - Integral Flanged Ends according to ASME B16.5

REGULAR PORT		1/4		3/8		1/2		3/4		1		1.1/4		1.1/2		2	
		mm	in.	mm	in.	mm	in.	mm	in.	mm	in.	mm	in.	mm	in.	mm	in.
Class 150 **F1-810**	A	-	-	-	-	108	4.25	117	4.64	127	5.00	-	-	165	6.49	178	7.00
Class 300 **F3-810**	A	-	-	-	-	140	5.51	152	6.02	165	6.49	-	-	190	7.51	216	8.50
Class 600 **F6-810**	A	-	-	-	-	165	6.49	190	7.51	216	8.50	-	-	241	9.48	292	11.5
Handwheel	B	-	-	-	-	80	3.14	80	3.14	110	4.33	-	-	130	5.11	130	5.11
Center to	Class 150/300	C	-	-	-	-	170	6.69	195	7.67	203	7.99	-	-	243	9.56	262	10.3
Top Open	Class 600	C	-	-	-	-	148	5.82	163	6.41	178	7.00	-	-	243	9.56	262	10.3
Dia. of Port		D	-	-	-	-	9.6	0.38	14	0.55	18	0.70	-	-	30	1.18	37	1.45
Approx. Weight	Class 150	Kg / Lb	-	-	-	-	3.4	7.5	3.8	8.3	5.7	12.5	-	-	9.7	21.4	13.2	29.1
	Class 300	Kg / Lb	-	-	-	-	3.9	8.6	5	11.0	6.2	13.6	-	-	12	26.4	16.5	36.3
	Class 600	Kg / Lb	-	-	-	-	4	8.8	5.2	11.4	75	16.5	-	-	15	33.0	20.5	45.1

End to End dimensions according to ASME B16.10
(*) End to end dimension according to ANSI B16.10.

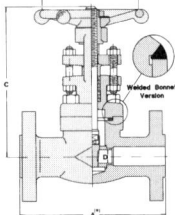

RATINGS: Carbon Steel
Class 150 - 285 p.s.i. @ 100°F
Class 300 - 740 p.s.i. @ 100°F
Class 600 - 1480 p.s.i. @ 100°F

CLASS 150-300-600
ROUND BOLTED BONNET - FULL PORT - ISO 15761
Outside Screw & Yoke - Integral Flanged Ends according to ASME B16.5

REGULAR PORT		1/4		3/8		1/2		3/4		1		1.1/4		1.1/2		2	
		mm	in.	mm	in.	mm	in.	mm	in.	mm	in.	mm	in.	mm	in.	mm	in.
Class 150 **F1-610**	A	-	-	-	-	108	4.25	117	4.64	127	5.00	-	-	165	6.49	178	7.00
Class 300 **F3-610**	A	-	-	-	-	140	5.51	152	6.02	165	6.49	-	-	190	7.51	216	8.50
Class 600 **F6-RJ610**	A	-	-	-	-	165	6.49	190	7.51	216	8.50	-	-	241	9.48	292	11.5
Handwheel	B	-	-	-	-	110	4.33	110	4.33	130	5.11	-	-	250	9.84	250	9.84
Center to	Class 150/300	C	-	-	-	-	170	6.69	195	7.67	210	8.26	-	-	262	10.3	327	12.8
Top Open	Class 600	C	-	-	-	-	244	9.60	268	10.5	310	12.2	-	-	391	15.4	430	16.9
Dia. of Port		D	-	-	-	-	14	0.55	18	0.70	24	0.94	-	-	37	1.45	48	1.89
Approx. Weight	Class 150	Kg / Lb	-	-	-	-	3.6	7.9	4.8	10.5	6.5	14.3	-	-	12	26.4	18	39.6
	Class 300	Kg / Lb	-	-	-	-	4.1	9.0	5.5	12.1	7.0	15.4	-	-	13	28.6	19	41.8
	Class 600	Kg / Lb	-	-	-	-	6	13.2	11	24.2	13	28.6	-	-	27	59.4	30	66.0

End to End dimensions according to ASME B16.10
Spiral wound gasket joint for #150 - #300
Ring-Joint gasket according to ASME B 16.20 - API 6A
(*) End to end dimension according to ANSI B16.10.

RATINGS: Carbon Steel
Class 1500 - 3705 p.s.i. @ 100°F

CLASS 1500
ROUND BOLTED BONNET RJ - FULL PORT - ISO 15761
Outside Screw & Yoke - Integral Flanged Ends according to ASME B 16.5

FULL PORT	F9-RJ910	1/4		3/8		1/2		3/4		1		1.1/4		1.1/2		2	
		mm	in.	mm	in.	mm	in.	mm	in.	mm	in.	mm	in.	mm	in.	mm	in.
End to End	A	-	-	-	-	216	8.50	229	9.01	254	10.0	-	-	305	12.0	368	14.5
Handwheel	B	-	-	-	-	110	4.33	130	5.11	130	5.11	-	-	250	9.84	300	11.8
Center to Top Open	C	-	-	-	-	260	10.2	300	11.8	300	11.8	-	-	390	15.3	420	16.5
Dia. of Port	D	-	-	-	-	14	0.55	18	0.70	24	0.94	-	-	37	1.45	48	1.89
Approx. Weight	Kg / Lb	-	-	-	-	11	24.2	16	35.2	19	41.8	-	-	35	77.1	59	130.0

End to End dimensions according to ASME B16.10
Spiral wound gasket joint available on request
Ring-Joint gasket according to ASME B 16.20 - API 6A

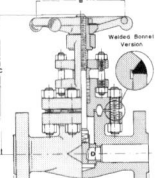

RATINGS: Carbon Steel
Class 2500 - 6170 p.s.i. @ 100°F

CLASS 2500
ROUND BOLTED BONNET RJ - FULL PORT - B16.34
Outside Screw & Yoke - Integral Flanged Ends according to ASME B 16.5

FULL PORT	F25-RJ2510	1/4		3/8		1/2		3/4		1		1.1/4		1.1/2		2	
		mm	in.	mm	in.	mm	in.	mm	in.	mm	in.	mm	in.	mm	in.	mm	in.
End to End	A	-	-	-	-	264	10.4	273	10.7	308	12.1	-	-	384	15.1	451	17.7
Handwheel	B	-	-	-	-	130	5.11	130	5.11	250	9.84	-	-	300	11.8	300	11.8
Center to Top Open	C	-	-	-	-	304	11.9	315	12.4	368	14.5	-	-	445	17.5	538	22.2
Dia. of Port	D	-	-	-	-	14	0.55	18	0.70	24	0.94	-	-	37	1.45	37	1.45
Approx. Weight	Kg / Lb	-	-	-	-	19	41.8	21	46.2	40	88.1	-	-	62	136.5	92	202.6

End to End dimensions according to ASME B16.10
Ring-Joint gasket according to ASME B 16.20 - API 6A

Figure 2–5 Gate valve, bolted bonnet, outside screw and yoke. (Printed with the kind permission of OMB Valves, spa, Italy.)

2.4 Valves 89

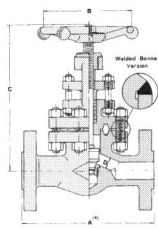

Figure 2–6 Globe valve, bolted bonnet, outside screw and yoke. (Printed with the kind permission of OMB Valves, spa, Italy.)

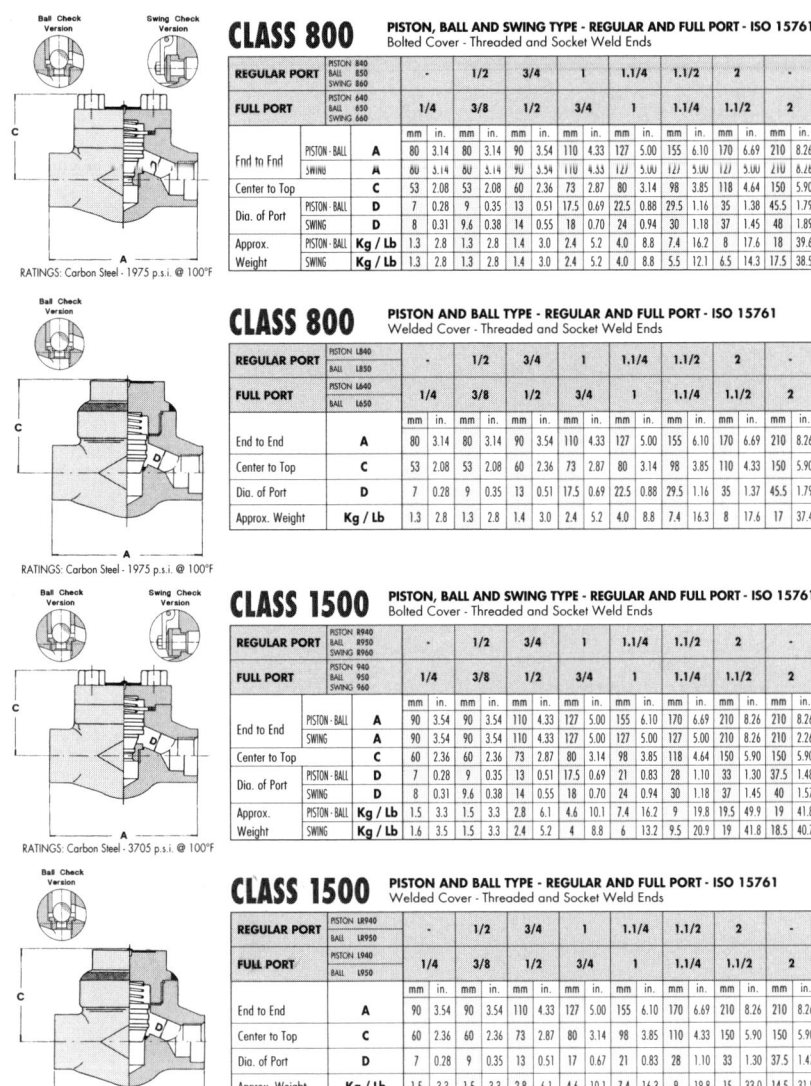

Figure 2–7 *Piston, ball and swing type threaded or socket weld ends. (Printed with the kind permission of OMB Valves, spa, Italy.)*

2.4 Valves

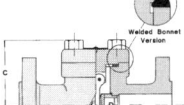

CLASS 150-300-600 — PISTON AND BALL TYPE - REGULAR PORT - ISO 15761
Bolted Cover - Integral Flanged Ends according to ASME B16.5

REGULAR PORT

			1/4		3/8		1/2		3/4		1		1.1/4		1.1/2		2	
			mm	in.	mm	in.	mm	in.	mm	in.	mm	in.	mm	in.	mm	in.	mm	in.
Class 150	PISTON F1-840 / BALL F1-850	A	-	-	-	-	108	4.25	117	4.64	127	5.00	-	-	165	6.49	203	7.99
Class 300	PISTON F3-840 / BALL F3-850	A	-	-	-	-	152	6.02	178	7.00	203	7.99	-	-	229	9.01	267	10.5
Class 600	PISTON F6-840 / BALL F6-850	A	-	-	-	-	165	6.49	190	7.51	216	8.50	-	-	241	9.48	292	11.5
Center to	Class 150	C	-	-	-	-	75	2.95	92	3.62	98	3.85	-	-	98	3.85	110	4.33
Top Open	Class 300-600	C	-	-	-	-	53	2.08	60	2.36	73	2.87	-	-	98	3.85	110	4.33
Dia. of Port		D	-	-	-	-	9	0.35	13	0.51	17.5	0.69	-	-	29.5	1.16	35	1.37
Approx. Weight	Class 150	Kg/Lb	-	-	-	-	2.9	6.4	3.2	7.0	4.3	9.5	-	-	6.5	14.3	14.5	31.9
	Class 300	Kg/Lb	-	-	-	-	3.6	7.9	4.2	9.2	6	13.2	-	-	12	26.4	16	35.2
	Class 600	Kg/Lb	-	-	-	-	4.1	9.0	4.7	10.4	6.3	13.8	-	-	13	28.6	17	37.4

End to End dimensions according to ASME B16.10

RATINGS: Carbon Steel
Class 150 - 285 p.s.i. @ 100°F
Class 300 - 740 p.s.i. @ 100°F
Class 600 - 1480 p.s.i. @ 100°F

CLASS 150-300-600 — SWING TYPE - REGULAR PORT - ISO 15761
Round Bolted Cover - Integral Flanged Ends according to ASME B16.5

FULL PORT

			1/4		3/8		1/2		3/4		1		1.1/4		1.1/2		2	
			mm	in.	mm	in.	mm	in.	mm	in.	mm	in.	mm	in.	mm	in.	mm	in.
Class 150 F1-860		A	-	-	-	-	108	4.25	117	4.64	127	5.00	-	-	165	6.49	203	7.99
Class 300 F3-860		A	-	-	-	-	152	6.02	178	7.00	216	8.50	-	-	241	9.48	267	10.5
Class 600 F6-860		A	-	-	-	-	165	6.49	190	7.51	216	8.50	-	-	241	9.48	292	11.5
Center to	Class 150	C	-	-	-	-	75	2.95	92	3.62	98	3.85	-	-	98	3.85	110	4.33
Top Open	Class 300-600	C	-	-	-	-	53	2.08	60	2.36	73	2.87	-	-	98	3.85	110	4.33
Dia. of Port		D	-	-	-	-	9.6	0.38	14	0.55	18	0.70	-	-	30	1.18	37	1.45
Approx. Weight	Class 150	Kg/Lb	-	-	-	-	2.9	6.4	3.2	7.0	4.3	9.5	-	-	6.5	14.3	14.5	31.9
	Class 300	Kg/Lb	-	-	-	-	3.6	7.9	4.2	9.2	6.1	13.4	-	-	13	28.6	16	35.2
	Class 600	Kg/Lb	-	-	-	-	4.1	9.0	4.7	10.4	6.3	13.8	-	-	13	28.6	17	37.4

End to End dimensions according to ASME B16.10

RATINGS: Carbon Steel
Class 150 - 285 p.s.i. @ 100°F
Class 300 - 740 p.s.i. @ 100°F
Class 600 - 1480 p.s.i. @ 100°F

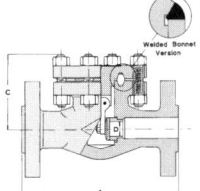

CLASS 150-300-600 — PISTON AND BALL TYPE - REGULAR PORT - ISO 15761
Bolted Cover - Integral Flanged Ends according to ASME B16.5

REGULAR PORT

			1/4		3/8		1/2		3/4		1		1.1/4		1.1/2		2	
			mm	in.	mm	in.	mm	in.	mm	in.	mm	in.	mm	in.	mm	in.	mm	in.
Class 150	PISTON F1-640 / BALL F1-650	A	-	-	-	-	108	4.25	117	4.64	127	5.00	-	-	165	6.49	203	7.99
Class 300	PISTON F3-RJ640 / BALL F3-RJ650	A	-	-	-	-	152	6.02	178	7.00	203	7.99	-	-	229	9.01	267	10.5
Class 600	PISTON F6-RJ640 / BALL F6-RJ650	A	-	-	-	-	165	6.49	190	7.51	216	8.50	-	-	241	9.48	292	11.5
Center to	Class 150	C	-	-	-	-	75	2.95	100	3.93	110	4.33	-	-	120	4.72	147	5.78
Top Open	Class 300-600	C	-	-	-	-	115	4.52	130	5.11	140	5.51	-	-	170	6.69	195	7.67
Dia. of Port		D	-	-	-	-	13	0.51	17.5	0.69	22.5	0.89	-	-	35	1.37	45.5	1.79
Approx. Weight	Class 150	Kg/Lb	-	-	-	-	3.2	7.0	3.5	7.7	4.6	10.1	-	-	7.0	15.4	16	35.2
	Class 300	Kg/Lb	-	-	-	-	4.6	10.1	6.1	13.4	9.1	20.0	-	-	16	35.2	21	46.2
	Class 600	Kg/Lb	-	-	-	-	4.8	10.5	6.3	13.8	9.3	20.5	-	-	16.5	36.3	22	48.4

End to End dimensions according to ASME B16.10
Spiral wound gasket joint and for #150
Ring-Joint gasket according to ASME B16.20 - API 6A

CLASS 150-300-600 — SWING TYPE - REGULAR PORT - ISO 15761
Bolted Cover - Integral Flanged Ends according to ASME B16.5

REGULAR PORT

		1/4		3/8		1/2		3/4		1		1.1/4		1.1/2		2	
		mm	in.	mm	in.	mm	in.	mm	in.	mm	in.	mm	in.	mm	in.	mm	in.
Class 150 F1-660	A	-	-	-	-	108	4.25	117	4.64	127	5.00	-	-	165	6.49	203	7.99
Class 300 F3-RJ660	A	-	-	-	-	152	6.02	178	7.00	216	8.50	-	-	241	9.48	267	10.5
Class 600 F6-RJ660	A	-	-	-	-	165	6.49	190	7.51	216	8.50	-	-	241	9.48	292	11.5
Center to	Class 150	C	-	-	-	-	75	2.95	100	3.93	110	4.33	-	-	120	4.72	147	5.78
Top Open	Class 300-600	C	-	-	-	-	115	4.52	130	5.11	140	5.51	-	-	170	6.69	195	7.67
Dia. of Port		D	-	-	-	-	14	0.55	18	0.70	24	0.94	-	-	37	1.45	48	1.89
Approx. Weight	Class 150	Kg/Lb	-	-	-	-	3.1	6.8	3.4	7.5	4.5	9.9	-	-	6.8	14.9	15.7	34.5
	Class 300	Kg/Lb	-	-	-	-	4.6	10.1	6.1	13.4	9.2	20.5	-	-	16.5	36.3	21	46.2
	Class 600	Kg/Lb	-	-	-	-	4.8	10.5	6.3	13.8	9.3	20.5	-	-	16.5	36.3	22	48.4

End to End dimensions according to ASME B16.10
Spiral wound gasket joint for #150
Ring-Joint gasket according to ASME B16.20 - API 6A

RATINGS: Carbon Steel
Class 150 - 285 p.s.i. @ 100°F
Class 300 - 740 p.s.i. @ 100°F
Class 600 - 1480 p.s.i. @ 100°F

Figure 2–8 *Piston, ball and swing type flanged ends. (Printed with the kind permission of OMB Valves, spa, Italy.)*

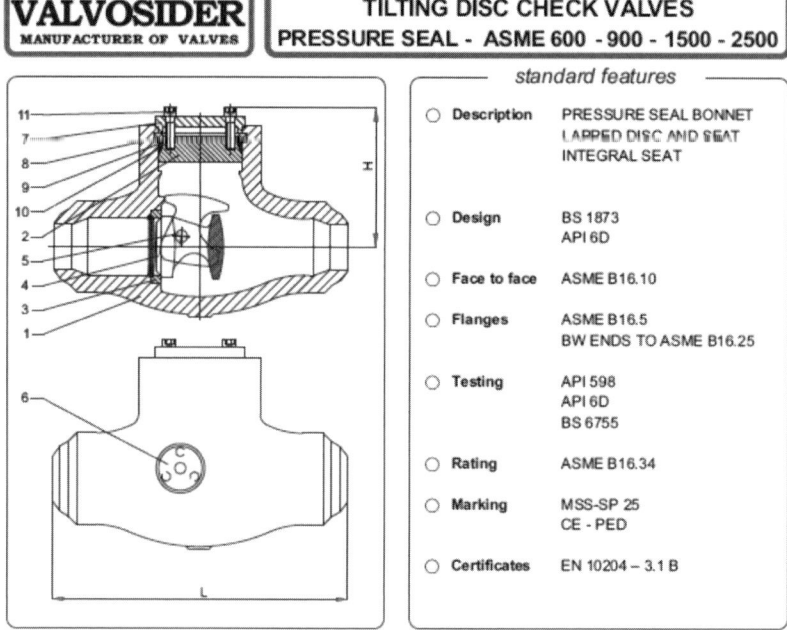

Figure 2–9 *Tilting-disc check valve, pressure seal bonnet. (Printed with the kind permission of Valvosider, srl, Italy.)*

2.4.3 Valve Classification

Valve Size

Valves are sized the same as pipe and pipe fittings, based on the nominal pipe size by U.S. customary units (NPS) in inches of their end connections, which are generally one of the following: threaded, socket weld, butt-weld, or flanged. In the metric system, valve size is designated by the *diamètre nominal* (DN) in millimeters. Many valves have a reduced internal port size; however, the valve size referenced is still based on the end connections. The term *small-bore valve* usually is applied to valves NPS 2 (DN 50) and smaller.

Pressure Class Ratings

Pressure-temperature ratings of valves are defined by ASME class numbers. Based on the material(s) of construction of the principle pressure-containing parts, the pressure-temperature ratings for each

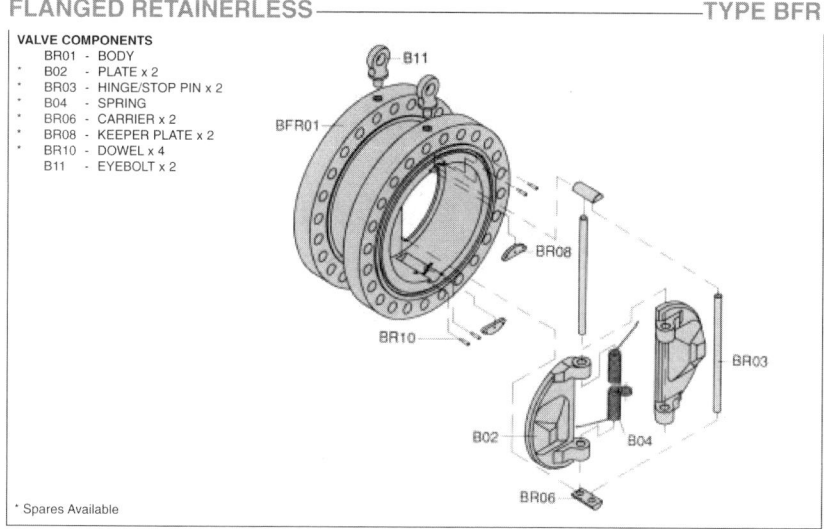

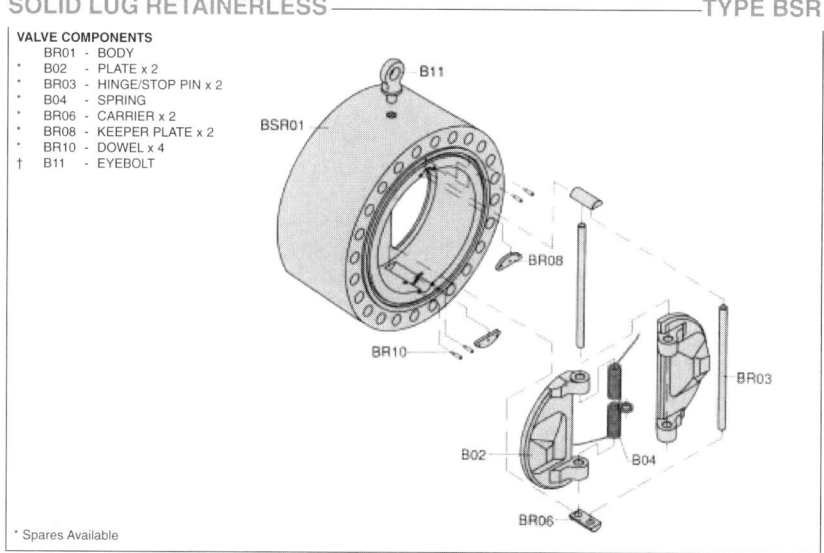

Figure 2–10 This is an exploded view of a wafer-type dual-plate check valve that will be held between two flanges: double flanged (top) and solid lug type (bottom). (Printed with the kind permission of Goodwin International, Ltd., United Kingdom.)

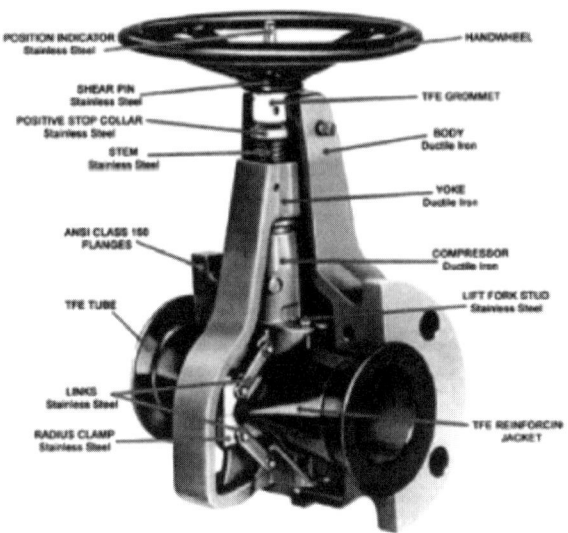

Figure 2–11 *Pinch valve with the components of construction. (Printed with the permission of Resistoflex.)*

class are tabulated to provide the maximum allowable working pressures, expressed as gauge pressures. These pressure-temperature classes are to be found in ASME B16.34.

The temperature shown for a corresponding pressure rating is the temperature of the pressure-containing shell or body of the component. These ratings apply to all valves regardless of type (see Appendix B, Figures B–5 through B–7).

ASME B16.34, Valves—Flanged, Threaded, and Welding End, is one of the most widely used valve standards. It defines three types of classes: standard, special, and limited. ASME B16.34 covers class 150, 300, 400, 600, 900, 1500, 2500, and 4500 valves. This standard is usually referenced in conjunction with other specific valve standards, such as API600, Steel Gate Valves; API602, Compact Steel Gate Valves; and API609, Lug- and Wafer-Type Butterfly Valves.

2.4.4 Valve Components

The components that make up the materials of construction of a valve fall into two distinct categories: pressure-containing parts and non-pressure-containing parts.

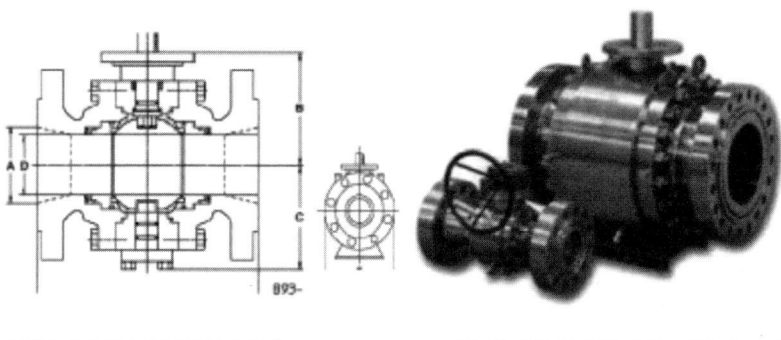

Figure 2–12 *Ball valve, flanged, side entry ASME class 150 full bore and reduced bore. (Printed with the kind permission of OMB Valves, spa, Italy.)*

Pressure Containing Parts

Valve body, bonnet or cover, disc, and body-bonnet bolting are classified as pressure-retaining parts of a valve and form the pressure envelope or boundaries of the valve. Not only must they have the mechanical strength for the design pressure and temperature conditions, but these

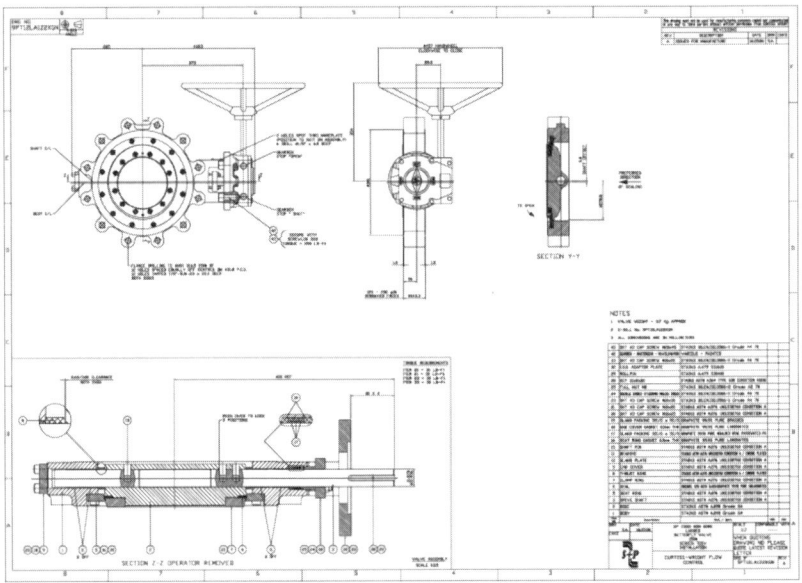

Figure 2–13   *Butterfly valve: (top) 84" class 150 butterfly valve for water service and (bottom) a general arrangement for a butterfly valve. (Printed with the permission of Curtiss Wright Controls.)*

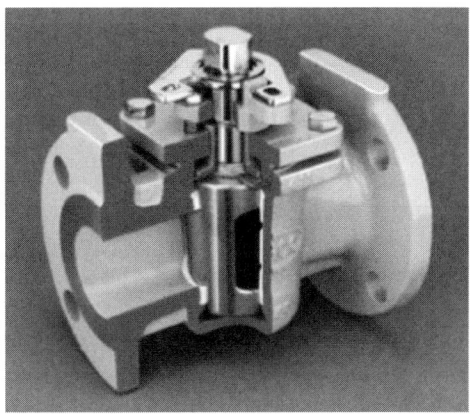

Figure 2–14 *Part section through a flanged plug valve. (Printed with the permission of Durco.)*

parts also come in contact with the process fluid and, therefore, must be compatible with the fluid and, if necessary, corrosion resistant for the lifetime of the plant or a predetermined timeframe. The following list provides a brief description of pressure retaining parts (see Appendix B, Figure B–9):

> *Body.* The valve body or shell forms part of the pressure containing envelope and is the essential framework that houses the internal valve parts. The body is in contact with the process media and should be compatible with the fluid that is transported. It has an inlet and an outlet, which can be threaded, flanged, or weld end. The body can either be of a cast or a forged construction.
>
> *Bonnet or Cover.* The bonnet or cover is connected to the valve body by flanges, threaded or welded to complete the pressure-retaining shell. This part is in contact with the process fluid. The body can be of either cast or a forged construction.
>
> *Bonnet or Cover Bolting.* This fastening assembly includes bolts, nuts, and occasionally washers. The bolting used must be made from materials acceptable for the application in accordance with the applicable code, standard, specification, or the governing regulation. It must be of a mechanical strength sufficient to create a leak-free joint between the valve

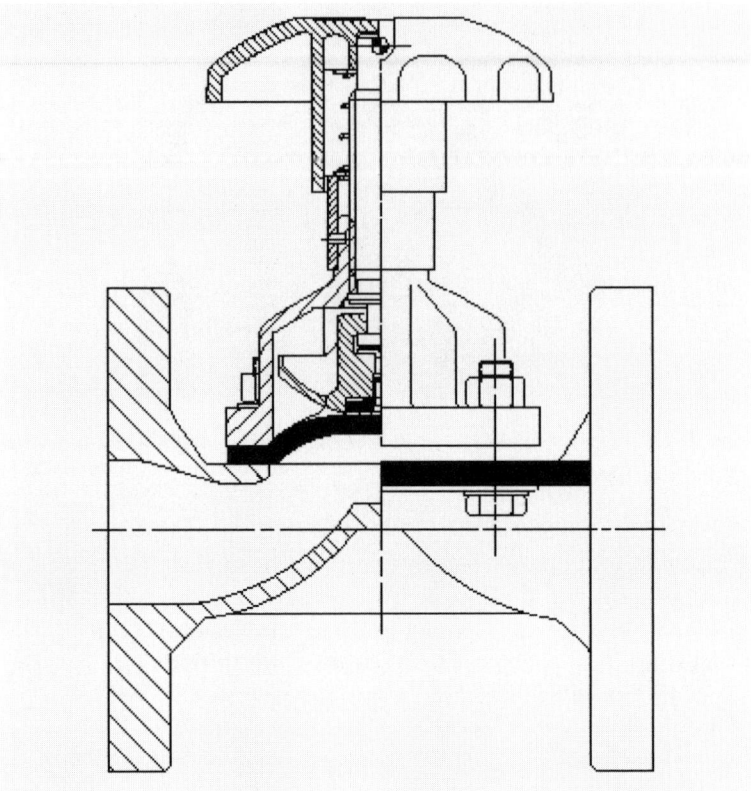

Figure 2–15 *Section through a rubber-lined flanged diaphragm valve. (Printed with the permission of Saunders.)*

body and bonnet; and because of this, it should be considered a pressure-containing piping component.

Body Bonnet Gasket. This component is trapped between the body and the bonnet. The gasket is a sealing element held in place by the compressive forces applied by the set of bolts.

Disc, Wedge, Ball, Plug, or Plate. Sometimes called the *closing element*, this part allows throttling or start/stop flow, depending on its position in relation to the seating/sealing surface. An intermediate position, between fully opened and fully closed, means that the part is in the throttling mode. The part is not permanently a pressure-retaining part. It falls into this classification only when the valve is fully closed;

when it is opened, the valve does not form part of the pressure-retaining envelope and, therefore, is not a pressure-containing part.

Non-Pressure-Containing Parts

These parts are not part of the pressure containing envelope, but they may be housed inside it. Non-pressure-retaining parts are the valve seat(s), stem, yoke, packing, gland bolting, bushings, hand wheel, and valve actuators:

> *Valve Seat(s).* A valve may have one or more sealing seats, and this surface isolates the fluid. Globe, butterfly, and swing-check valves usually are referred to as *single-seat valves*. Gate and ball valves could be either single- or double-seat valves. A gate valve has two seating surfaces, one on the upstream side and the other on the downstream side. The gate-valve disc or wedge has two seating surfaces, one on either side of the gate that comes in contact with the valve seats to form a seal for stopping the flow. The flow direction dictates that the downstream seat is more effective because of the force applied by the fluid. The downstream force makes the stem flex slightly and forces the gate against the downstream seat. Generally, a gate valve has a metal-to-metal sealing surface, which makes a leaktight joint more difficult; therefore, a certain degree of leakage is acceptable, and this is defined in a valve standard, such as ASME B16.34 or API 6D (see Appendix B, Figure B–1).
>
> *Valve Stem.* The valve stem is the part that applies the necessary torque to raise, lower, or rotate the closure element; it opens, closes, or positions the closure element. In the globe valve, this is a linear motion. For ball, plug, and butterfly valves, this is a rotary motion. The stem must be of sufficient mechanical strength not to shear during operation, and it is partially in contact with the process fluid, so the two must be compatible. There are several different types of stem valves:
>
> - *Rising Stem with Outside Screw and Yoke.* The part of the stem exposed to the outside environment is threaded, while the section of stem inside the valve is smooth. The stem threads on the outside, therefore, are isolated from

the process flow by the stem packing. There are two styles of stems, one with the handwheel fixed to the top of the stem, so that they rise and fall together, and the other with a threaded sleeve that causes the stem to rise through the center of handwheel. In the latter, a rising stem with outside screw and yoke (O.S.&Y.) is a common design for NPS 2 (DN 50) and larger valves.

- *Rising Stem with Inside Screw.* The threaded part of the stem is inside the valve body, and the stem packing is along the smooth section exposed to the outside atmosphere. In this case, the stem threads are in contact with the flow medium. When rotated, the stem and the handwheel rise together to open the valve. This design is commonly used in the smaller-sized low- to moderate-pressure gate, globe, and angle valves.

- *Nonrising Stem with Inside Screw.* The threaded section of the stem is inside the valve and does not rise. The valve disc travels along the stem like a nut when the stem is rotated. Stem threads are exposed to the flow medium and, as such, are subjected to its impact. Therefore, this design is used where space is limited to allow linear stem movement, and the flow medium does not cause erosion, corrosion, or wear and tear of stem material.

- *Sliding Stem.* The stem does not rotate, and it is without a thread. It slides in and out of the valve packing to close, open, or position the valve closure member. This design is used in hand-lever-operated, quick-opening valves. It is also used in control valves operated by hydraulic or pneumatic cylinders.

- *Rotary Stem.* This is the most commonly used stem design in ball, plug, and butterfly valves. A quarter-turn motion, (90° rotation of the stem) opens, closes, or positions the valve closure member.

Stem Packing. The stem packing of a valve performs one or both of the following functions, depending on the application: prevents leakage of flow medium to the environment (most common) or prevents outside air from entering the valve in vacuum applications (less common). The stem packing is contained in a part called the *stuffing box*, usually in a series of rings compressed in place by a gland and bolting to achieve a good seal. The stem packing must have the mechanical characteristic to be compressed and create a sealing contact against the walls of the chamber of the stuffing box. The packing also is partially in contact with the process fluid, so the two must be compatible.

Stem Protector. A stem protector is used when the gate or globe valve is of an outside-screw-and-yoke, rising-stem design.

Backseat. The backseat is the part with a shoulder on the stem and a mating surface on the underside of the bonnet. This combination forms a seal when the stem is in the fully open position. It prevents leakage of flow medium from the valve shell into the packing chamber and, consequently, to the environment.

Yoke. A yoke has "arms" that connects the valve body or bonnet with the actuating mechanism. In some cases, it provides support for the gland pull-down bolts. The yoke must be mechanically robust enough to withstand forces, moments, and torque developed by an actuator.

Yoke Bushings. The yoke bushings are internally threaded nuts held in the top of a yoke through which the valve stem passes.

2.5 Bolts and Gaskets (Fasteners and Sealing Elements)

Potential leak paths in a piping system should be avoided wherever possible; however, there are times when their presence is necessary for erection and maintenance reasons. As long as the correct materials are selected, the bolting method and procedures necessary to create a leak-free joint are in place, and suitably qualified personnel are available, then a leak-free joint can be achieved.

This section deals with mechanical bolted flanged joints. It covers the necessary jointing components, gaskets and bolts, the various materials of construction, and the procedures necessary to complete a leak-free seal between the two compatible flange faces.

The bolted flanged joint is potentially the weakest link in a piping system, and it is essential that the design limitations of the flange type and the materials of construction are never exceeded. A flanged connection can be assembled and disassembled more easily than a welded joint, which should be considered an advantage if the mechanical joint is leak free.

A number of international standards cover the individual components required for a bolted jointing: flanges, gaskets, and bolts that must be used to achieve a satisfactory joint. Several standards have been written to enable designers to design bolted joints that ensure mechanical integrity, and they must be followed to obtain the best results.

2.5.1 The Process of Joint Integrity

A leak-free mechanical joint, between a flange face, gasket, and bolting is achievable if the following areas and appropriate international standards are carefully considered:

- Pressure and temperature design conditions of the internal fluid.
- External environmental conditions.
- Flange face design.
- Flange material.
- Gasket type and materials of construction.
- Fastener (bolt and nut) material.
- Bolting lubricant.
- Bolting procedure—torque tensioning and bolt-up sequence.
- Skilled workforce.

Failure to address all of these will likely lead to a leak path that may result in a costly plant shutdown.

2.5.2 Flange Joint Components

Flanges

As discussed earlier, for a plant designed to one of the ASME B31 codes, several flange types are available for the erection of process piping systems. For standardization and interchangeability, these various options are covered in numerous international standards. ASME has its own group of standards that cover the relevant components used in a mechanical bolt-up.

Flange Standards

A variety of standards are used in the design and selection of flanges. The following codes and standards relate to pipe flanges, and these are the ones most commonly used for process piping systems:

    ASME Codes and Standards

        B16.1, Cast Iron Flanges and Flanged Fittings.

        B16.5, Pipe Flanges and Flanged Fittings.

        B16.24, Bronze Flanges and Fittings—150 and 300 Classes.

        B16.42, Ductile Iron Pipe Flanges and Flanged Fittings—150 and 300 Classes.

        B16.47, Large Diameter Steel Flanges.

    Section VIII

        Division 1, Pressure Vessels.

        Appendix 3, Mandatory Rules for Bolted Flange Connections.

    API Specification

        Spec 6A, Specification for Wellhead and Christmas Tree Equipment.

The three most commonly used flange standards for process and utilities pipework are ASME B16.5, ASME B16.47, and API 6A, which specify flanges for wellhead and Christmas tree equipment.

End Connection of Flange

The end connection of the flange specifies how the flange is attached to a neighboring pipe. There are several alternatives, each with its own technical and commercial merits. This is a list of commonly available flange end types that have been discussed earlier: weld neck, slip on, lap joint, threaded, and socket weld.

Flange Faces

Three types of flange faces are commonly found. The surface finish of the faces is specified in the flange standards quoted previously:

> *Raised Face.* The raised face is the most commonly used facing employed for steel flanges. The facing on the RF flange has a concentric or a spiral phonographic groove with a controlled surface finish. Sealing is achieved by compressing a flat, soft, or semi-metallic gasket between mating flanges in contact with the raised face portion of the flange.
>
> *Ring-Type Joint.* This type is typically used for more severe duties than the RF surface, usually ASME classes above 900; however, it is valuable in the lower-pressure classes.
>
> - *RTJ for API 6A Type 6B, ASME B16.5 Flanges.* The seal is made by plastic deformation of the metallic RTJ gasket into the groove on the flange face, resulting in intimate metal-to-metal contact between the gasket and the flange groove. The faces of the two opposing flange faces do not come into direct contact with each other, because a gap is maintained by the presence of the gasket. Such RTJ flanges normally have raised faces, but flat faces may also be used or specified.
>
> - *RTJ for API 6A Type 6BX Flanges.* API 6A Type 6BX flanges have raised faces. These flanges incorporate special metallic ring joint gaskets. The pitch diameter of the ring is slightly greater than the pitch diameter of the flange groove. This factor preloads the gasket and creates a

pressure-energized seal. A Type 6BX flange joint that does not achieve face-to-face contact will not seal and, therefore, must not be put into service.

Flat Face. Flat-face flanges are a variant of raised-face flanges. Sealing is achieved by compression of a flat nonmetallic gasket between the two serrated surfaces of the mating FF flanges. The gasket covers the entire face of the flange sealing surface. FF flanges are normally used for the least arduous duties, such as low-pressure water piping having class 125 and class 250 flanges and flanged valves and fittings.

Less Commonly Used Flange Faces. Other alternative types of flanges are available; however, due to international standardization in the energy industry, they are very rarely used on projects designed to one of the ASME B31 codes.

- *Male-and-Female Facings.* The outer diameter of the female face acts to locate and retain the gasket. Custom male-and-female facings are commonly found on the heat exchanger shell to channel and cover flanges.

- *Tongue-and-Groove Facings.* Tongue-and-groove facings are standardized in both large and small types. They differ from male-and-female facings in that the inside diameters of the tongue-and-groove do not extend into the flange base, thus retaining the gasket on its inner and outer diameter. These are commonly found on pump covers and valve bonnets.

Flange Specification

A flange is specified by identifying the following information:

Type and Facing (For Example, WN/RTJ, SO/RF, Lap Joint/RF). This is a very short identifier that describes the design of the flange and the type of flange facing.

Nominal Pipe Size. NPS is a dimensionless designation to define the nominal pipe size of the connecting pipe, fitting, or nozzle. Examples include NPS 4 and NPS 6.

Flange Pressure Class. This designates the pressure temperature rating of the flange, which is required for all flanges and taken from ASME B16.5. Examples include ASME Classes 150, 300, 600, 900, 1500, and 2500.

Standard. Flange dimensions and material group from ASME B16.5.

Material. A material specification for flanges must be specified and be compatible to the piping material specifications.

Pipe Schedule. This is only for WN, composite lap-joint, and swivel-ring flanges, where the flange bore must match that of the pipe, such as schedule 40, 80, 120, or 160.

Gaskets

A gasket is a sealing element placed between the two flange faces and held in position by the compressive forces of the set of bolts located around the circumference of flange blades. Gaskets are constructed from a variety of materials and, in some cases, a combination of materials. The gasket must be capable of maintaining a leak-free seal during the lifetime of the joint at the design pressures and temperatures of the fluid being transported.

Gasket Standards

The following are three of the most commonly used international standards for gaskets specified in process plants designed to ASME B31 codes. They include dimensions, tolerances, fabrication, and marking of gaskets:

ASME B16.20, Metallic Gaskets for Pipe Flanges, Ring Joint, Spiral Wound, and Jacketed.

ASME B16.21, Nonmetallic Flat Gaskets for Pipe Flanges.

API 6A, Specification for Wellhead and Christmas Tree Equipment.

Types of Gaskets

Gaskets can be divided into three categories based on their materials of construction: nonmetallic, semi-metallic, and metallic.

Nonmetallic Gaskets

Nonmetallic gaskets are cut from flat, soft sheet; and they are used with flat-face mating flanges, generally for low-pressure class applications (ASME 150) and very rarely specified for pressures above 20 bag. Prior to withdrawal from the industry because of health and safety issues, asbestos was commonly used and termed *compressed asbestos fiber* (CAF) gaskets. The term has been changed and now is referred to as *compressed nonasbestos fiber* (CNAF) gaskets.

Nonmetallic gaskets are made from materials such as elastomers (natural and synthetic rubbers), Teflon (PTFE), and flexible graphite. Full-face gasket types are held between flat-face flanges. Flat-ring gasket types, which do not cover the full face and are located within the bolt circle of the flange, can be used with raised-faced flanges.

Flat metallic gaskets are available in a variety of steels, copper, and other materials; however, they are very rarely used on oil and gas projects.

Gasket dimensions are covered in ASME B16.5 up to NPS 24" and in ASME B16.47 Series A and B for NPS 26" and above.

Semi-Metallic Gaskets

Semi-metallic gaskets are a combination of two or more metallic and nonmetallic materials. The metal gives strength and robustness to the gasket, and the nonmetallic portion of a gasket provides sealability to the integrated component. These composite gaskets are used from low-pressure (ASME 150) classes through to high-pressure classes (ASME 2500), although ring-type joint gaskets tend to be more commonly used at classes ASME 1500 and above. Semi-metallic gaskets are spiral wound, metal jacketed, camprofile/Kamprofile, and metal-reinforced graphite gaskets. Spiral wound gaskets are the most commonly used gaskets for raised-face flanges. They are used in all pressure classes from ASME class 150 to class 2500. The section of the gasket that creates the seal between the flange faces is the spiral wound section, which houses the soft sealing element. It is manufactured by winding a preformed metal strip and a soft filler material around a metal mandrel. The spiral wound gasket has an outer or centering ring that holds the windings in place and locates the gasket within the bolt circle. For special services and at higher pressure ratings, it can have an additional inner ring that helps to secure the winding and prevents it from buckling inward and entering the process flow. Spiral wound gaskets are manufactured to ASME B16.21 standard to suit flanges designed to ASME B16.5 and ASME B16.47.

The term *camprofile* or *Kammprofile gaskets* comes from the Dutch word for comb, Kam, which describes the design of the gasket. The gaskets are made from a solid serrated metal core faced on each side with a soft nonmetallic material. Kammprofile gaskets are used on all pressure classes from class 150 to class 2500 in a wide variety of service fluids and operating temperatures.

Jacketed gaskets are made from a nonmetallic gasket material housed within a metallic jacket. This inexpensive gasket arrangement is used occasionally on standard flange assemblies, valves, and pumps. Jacketed gaskets are easily fabricated in a variety of sizes and shapes and provide an inexpensive gasket for heat exchangers, shell, channel, and cover flange joints. Their metal seal makes them unforgiving to irregular flange finishes and cyclic operating conditions.

Metallic gaskets are usually constructed from one grade of metal to a predetermined size and shape. The most widely used type of metallic gasket in the process industry is the ring-type joint, which can be used at elevated pressures and temperatures.

Ring-joint gaskets, are manufactured to ASME B16.20 to suit flanges designed to ASME B16.5 and ASME B16.47. The API 6A standard covers both flanges and matching gaskets, which can be divided into three distinct groups:

- Style R, either ring or octagonal.

- Style RX, a pressure-energized adaptation of the standard style R ring-joint gasket.

- Style BX, pressure-energized ring joints designed for use on pressurized systems up to 20,000 psi (138 MPa).

Flange faces using BX-style gaskets come in contact with each other when the gasket is correctly fitted and bolted up. The BX gasket incorporates a pressure-balance hole to ensure equalization of pressure, which may be trapped in the grooves.

Function of Gaskets

The primary function of any flanged assembly is to create a leak-free joint during the projected lifetime of the process piping system. The gasket is the sealing element in the assembly, and its purpose is to deform into the spirally grooved surface of the two flange faces to create a leak-free seal that prevents the fluid from leaking out and the ingress of the outside environment.

The leak performance of the gasket depends on the loads applied on the gasket during the bolt-up procedure. The higher the gasket stress, the higher the leaktightness capability. Having said this, it is also very important not to overtighten the bolts and "squeeze" the gasket between the two flanges. This is why it is essential that the correct bolting procedure is followed. Correct flange and gasket selection are meaningless if the bolting procedure is ignored.

There have been cases where, in error, a gasket has not been installed between two flanges, but because the bolting procedure was correct, the flanged assembly passed a low-pressure test with no leakage.

Gasket Selection

The type of gasket and its materials of construction depend on the flange facing, service, design pressure and temperature, and external environment.

Bolts and Nuts

To complete any flanged assembly, two additional components are essential: bolts (stud or machine) and nuts. Bolts and nuts are the fasteners that provide compressive clamping forces to trap the gasket between the two sealing surfaces on the flange faces. The term *bolting* applies to the bolt, nuts, and if required, the washer.

The ASME pressure class and the NPS size of flange determine the number of bolts, the outside diameter and length of the bolt, and the geometrical positioning of the bolt circle on the flange. The number of bolt holes increases by four: 4, 8, 12, 16, 20, 24, and so forth. This bolting pattern has been carefully calculated to create a leak-free joint plus an acceptable safety factor. For example, NPS 24", ASME class 300 flanges require 24 bolts in the bolt pattern. If the correct bolting procedure is applied, this would probably result in a leak-free join with 20 bolts, because of the safety margin.

The following international standards pertain to bolting:

>ASME B1.1, Unified Inch Screw Threads.
>
>ASME B18.2.1, Square and Hex Bolts and Screws.
>
>ASME B18.2.2, Square and Hex Nuts.
>
>ASME B18.21.1, Lock Washers.
>
>ASME B18.22.1, Plain Washers.
>
>ASTM F436, Mechanical Properties of Plain Washers.

Bolts

A bolt is a steel fastener made from a bar, with an integral head at one end and a shank length that is threaded. Stud bolts usually are used for flanges in a process plant. These are bars that are partially or totally threaded along the length. Coupled with the stud bolt are two hexagonal nuts, which are tightened to compress the gasket and create a leak-free seal. Bolts can be tightened either manually, using a torque wrench, or, if a more accurate method is required, using a hydraulic stud tension meter that delivers a preselected load.

Nuts

Heavy-series hexagonal nuts generally are used with studs on pressure piping. The nonbearing face of a nut has a 30° chamfer, while its bearing face is finished with a washer face.

Bolt and Nut Selection

Bolts and nuts should be selected to conform to the design specifications set out with the flange design. Care is taken to ensure that the correct grade of material is selected to suit the recommended bolting temperature and stress ranges. The following information should be specified when ordering bolts and nuts:

1. Quantity.
2. Grade of material, identifying symbol of bolt or nut.
3. Form—bolts or stud bolts; nuts, in regular or heavy series.
4. Dimensions—nominal diameter and length; diameter of plain and reduced portion, length of thread (if applicable).
5. Identification of tests in addition to those stated in the standard.
6. Manufacturer's test certificate (if required). Fully threaded stud bolts and heavy series nuts are most common in industrial applications.

Function of Bolts

The function of a bolt is to provide a clamp load that compresses the gasket between the two flange faces and creates a leak-free seal. As the load is applied to the bolt, the nut travels down the shank and a compressive stop on the back of the flange face stretches the bolt and compresses the gasket to create a seal.

2.5.3 The Flanged Joint System

The individual components of flanges, gaskets, nuts, and bolts work together in unison, and combined with the correct lubrication and bolt-up procedure, create a seal that prevents external leakage.

Flange Condition

The condition of flange surfaces and selection of the proper flange material play very important parts in achieving a leak-free joint assembly. It is essential that the following flange conditions are within acceptable limits: surface finish, waviness or roughness, flatness, surface imperfections, and parallelism.

Gasket Condition

The gasket must be new: It should never be reused. It must be clean and clear of all visible surface defects. Imperfections and foreign material may create a radial leak path that will affect the sealing capabilities of the gasket.

Gasket Quality

The gasket must be procured from a reputable manufacturer with support data sheets that specify the material of construction, design, and dimensional standards.

Bolt Condition

Bolts and nuts may be reused: however, it is always best to use new components. Ensure bolts and nuts are clean, free of rust, and the nut runs freely on the bolt threads. Install bolts and nuts well lubricated by using a high-quality antiseize lubricant to the stud threads and the nut face.

Methods of Bolt Tightening

Once the total bolt loads (W) are calculated for the flanges, specifications and procedures should be adopted outlining how to achieve the design bolt load. The total bolt load (W) for the flange is divided by the number of bolts to determine the individual bolt preload (Fp). To achieve improved leak tightness, sufficient and uniform gasket stress must be realized in the field. This obviously requires uniform and correctly applied bolt load. The higher the requirement to reduce leakage, the more controlled must be the method bolt tightening.

The common methods of bolt tightening use hammer or impact wrenches, torque wrenches, and hydraulic tensioning systems. Each method has its merits.

Hammer or Impact Wrenches Method

This method remains the most common form of bolt tightening. The advantages are speed and ease of use. Disadvantages include a lack of preload control and the inability to generate sufficient preload on large bolts.

Torque Method

Torque wrenches are often regarded as a means to improve control over boltpreload in comparison with hammer-tightening methods.

Assembly

The leak-free characteristic of a bolted joint does not depend solely on the gasket itself but on a combination of variables, many of which are outside the control of the gasket manufacturer. In most cases, leakage is not due to gasket failure but is more likely to result from poor installation, assembly, or bolting practices; damaged flanges; incorrect or no bolt lubricant; or a combination of variables associated with a bolted gasketed flanged joint.

The following must be considered when installing a gasket:

Fasteners (nuts and bolts). The correct material for the nut and bolt must be selected—type, proper material, grade, appropriate coating or plating, class, correct stud and bolt length.

Assembly. A bolt-up procedure must be very carefully followed to achieve a leak-free seal.

- Install a new gasket on the gasket seating surface and bring the mating flange in contact with the gasket.

- Do not apply any compounds on the gasket or gasket seating surfaces.

- Install all bolts, making sure that they are free of any foreign matter and well lubricated. Lubricate nut bearing surfaces as well (lubrication is not required for PTFE coated fasteners).

- Run-up all nuts finger tight.

- Develop the required bolt stress or torque incrementally in a minimum of four steps in a crisscross pattern. The initial prestress should be no more than 30% of the final required bolt stress. After following this sequence, a final tightening should be performed bolt to bolt to ensure that all bolts have been evenly stressed. Note: The use of hardened washers enhances the joint assembly by reducing the friction due to possible galling of the nut bearing surfaces.

CHAPTER 3

Metallic Materials for Piping Components

Technological advances and the use of computer aided design has had no direct effect on the materials of construction of piping components for process systems. Carbon steel is the workhorse of industry, and coupled with an adequate corrosion allowance, this material can cover most eventualities. Low-temperature carbon steel is used for subzero temperatures and low-alloy carbon steel is used at elevated temperatures. After carbon steel, stainless steel is the next most used metal, followed by the duplexes and more exotic metals. Very little has changed over the last 30 years. The purpose of this chapter is not to go into metallurgy in depth but to cover the most relevant points that apply to the base materials of construction of piping components.

The construction material for process piping still is dominated by the use of carbon steel, low-alloy carbon steel, low-temperature carbon steel, and supported by the numerous stainless steels grades. Used to a lesser degree are the exotic materials, such as alloy 625 and alloy 825, and nonmetallic materials, like GRP and PVC.

The selection of material to be used within a piping system is the responsibility of the metallurgist/corrosion engineer, who creates a material selection report that identifies what base material must be used in both process and utility systems. This MSR forms the basis from which the piping material classes are created to cover the numerous fluids at various pressures and temperatures within the process plant. Piping engineers do not necessarily have to know the fine details of this specialist field; however, they should be aware why the metallurgist came to his or her conclusion.

When selecting the materials of construction for a piping system to be used within a project designed to ASME B31 code, numerous factors have to be considered. The material selected must be suitable for the following conditions:

- Media (corrosion, erosion).
- Design life (usually 20 to 25 years).
- Design temperature range.
- Design pressure range.
- External environment (aboveground, buried, subsea, etc.).
- Mechanical cycling.
- Thermal cycling.
- Mechanical impact.
- Method of jointing.
- Economics.
- Availability of material.
- Labor available.

The most commonly used metallic piping materials are listed in ASME B31.3; however, materials outside of this list can be used as long as they are supported by the appropriate data sheets and independent testing reports. All materials have different chemical compositions, which have an effect on the mechanical and physical characteristics and their resistance to corrosion at differing temperatures and pressures.

The most commonly used material in the oil and gas industry is carbon steel, which performs satisfactorily at temperatures between –29°C and 427°C, and it can be used at the highest ASME and API pressure ratings.

3.1 Properties of Piping Materials

The mechanical and physical characteristics of a particular material, at a given temperature and pressure, can be predicted if the following are known: chemical composition, method of manufacture, and additional heat treatment. These factors determine how a particular metal behaves under certain conditions, and they enable the engineer responsible for material selection to specify material with confidence. This knowledge must go right down to the atomic construction of a material.

3.1.1 Chemical Properties of Metals

The chemical composition of a material usually is measured by the relative atomic weight percent of the various elements (metals or nonmetals) or compounds within the material.

Metals are rarely, if ever, used in their purest form, but have numerous elements, metallic and nonmetallic, added to create the desired mechanical characteristics and resistance to corrosion (see Appendix B, Figure B–3). Some of these are intentional additions and some are unavoidable and have to be controlled. This action, called *alloying*, improves and modifies the behavior of the metal. The most commonly used additional element to iron is carbon, which increases ultimate strength and hardness and lowers the ductility.

Elements also can be added to improve the material's machinability, formability, and weldability, which improves the manufacturing and fabrication process as the material is formed into its final dimensional shape as a piping component. Some alloying elements are listed later in this chapter, indicating what improvements they bring to steel.

Carbon steel is the most commonly used of all the construction materials for piping components; and it always contains the elements carbon, manganese, phosphorous, sulfur, and silicon in varying percentages, depending on the grade. Small amounts of other elements may be found either entering as gases during the steelmaking process (hydrogen, oxygen, nitrogen) or introduced intentionally through the ores or metal scrap used to make the steel (nickel, copper, molybdenum, chromium, tin, antimony, etc.). The addition of each element has a specific effect on the properties of the steel.

A great deal of research has been carried out during the development of the metals used in piping design and construction; and the data collected means that, if the specifications are adhered to, then the behavior of the metal is predictable, which gives the specifying engineer confidence.

The chemical composition, various grades, and the subsequent differing mechanical and physical properties of a metal are documented in various specifications, such as those from the ASME, API, and ASTM. The test methods necessary to record and document these properties are defined in ASTM specification E series.

3.1.2 Mechanical Properties of Metals

Mechanical properties of a metal are defined by the chemical composition, method of manufacture, and additional heat treatment. This information is essential when selecting a material for process piping.

The test methods for measuring these properties are covered in ASTM specification E and fall into two categories, strength and ductility. Listed next are the values that are covered in the numerous ASTMs, which are the most commonly specified material for ASME B31 code projects:

- Ultimate tensile strength, ksi or MPa.
- Yield, ksi or MPa.
- Elongation.
- Hardness, Brinell or Rockwell number.

Ultimate Tensile Strength

When a load is applied to a test piece, it will stretch as the loss of load-carrying cross section causes a reduction in the cross section until eventually it fractures and fails. The *ultimate tensile strength* (UTS) is defined as the maximum applied load divided by the original specimen cross-sectional area.

Yield Strength

When a test piece has a load applied beyond the point where elastic behavior can be maintained, the specimen begins to deform (yield) into a plastic state. In most cases, materials do not suddenly transform from an elastic to a plastic state. A gradual transition phase occurs, and this can be represented by a curve, or knee, in the stress-strain curve. There is no accurately defined transition between the elastic and plastic phases: however, a number of ASTM testing methods determine the yield strength of metals.

This yield test is performed at a constant rate of strain and is measured in newtons per square meter in the metric system or pounds per square inch of cross section in U.S. customary units.

Modulus of Elasticity (Young's Modulus)

The *modulus of elasticity* is defined as the ratio of normal stress to corresponding strain for tensile or compressive stresses. The material behavior in this range is elastic, and if the applied load is released, the material returns to its original shape.

The value of the slope in the elastic range is defined as *Young's modulus*. The modulus of elasticity is measured using the tension test, which consists of applying a load gradually increasing in either tension or compression, in a testing machine, to a specially prepared test piece.

3.1.3 Elongation and Reduction of Area

The ductility of the test piece can be established by measuring its length and minimum diameter before and after testing. Stretch of the specimen is represented as a percent elongation in a given length, usually 2" or 8". The diameter of the test piece decreases, or *necks down*, in ductile materials.

Hardness

Hardness is the ability of a material to resist deformation, which is determined by a standard test where the surface resistance to indentation is measured. The most commonly used hardness tests are defined by the shape or type of indent, the size, and the amount of load applied. The hardness numbers referenced constitute a nondimensioned, arbitrary scale, with increasing numbers representing harder surfaces. The two most referenced hardness test methods are Brinell hardness and Rockwell hardness, each one having a dedicated test machine with its own unique hardness scales. Hardness correlates approximately with ultimate tensile strength in metals. Hardness conversion numbers for commonly used material types can be found in ASTM Specification E140.

Toughness

Lack of toughness occurs when a material type fractures and exhibits little ductility in the locality of the break. This occurs in certain metals when a load is rapidly applied, and the ability of a material to resist such a brittle fracture is a measurement of its toughness. Materials with high ductility show toughness across a full temperature range. Other materials possess a level of toughness dependent on the metal temperature when the load is applied. In these metals, a transition from brittle to ductile behavior occurs over a narrow range of temperatures.

The two most used methods to measure the toughness of metal are the Charpy impact test, which is detailed in ASTM specification E 23, and the drop-weight test, which is covered in ASTM E 208.

The Charpy test requires a small, specially prepared test piece with a machined notch that is struck by a pendulum of a predetermined weight. The energy loss to the pendulum as it passes through and breaks the test piece, measured in kilojoules or ft/lb of force, is a measure of the toughness of the test piece.

The drop-weight test uses a very similar principle, but it requires a much larger test piece with a brittle, notched weld bead used as the crack starter. A predetermined weight is dropped from a set height onto the test piece, which is at a desired test temperature.

3.1.4 Physical Properties of Metals

Other physical properties of metals show characteristics that relate to the physics of the metal. Significant physical properties to the materials and design engineer are the density, thermal conductivity, thermal expansion, and specific heat, all of which are important factors during material selection.

>*Density.* Density is defined as the ratio of the mass of a material to its volume.
>
>*Thermal Conductivity.* The thermal conductivity of a material is its ability to transmit energy in the form of heat from a high-temperature source to a point of lower temperature. This characteristic to transmit heat usually is expressed as a coefficient of thermal conductivity (k), whose units are the quantity of heat transmitted through a known unit thickness per unit time per unit area per unit difference in temperature.
>
>*Thermal Expansion.* Thermal expansion is the ratio of the change in length per degree of temperature to a length at a given standard temperature. The coefficient units are length of growth per unit length per degree of temperature. The value of the coefficient varies with temperature.
>
>*Specific Heat.* Specific heat is a measure of the quantity of heat required to raise a unit weight of a material 1° in temperature.

In addition to the mechanical and physical properties already discussed are other characteristics of metals that have an important effect on the selection of material for process of piping components:

>*Grain Size.* When a metal solidifies from the molten state, it takes on a crystalline shape. A metal comprises a multitude of small crystals that combine to form a shape of the piece. These individual crystals are known as *grains*, and their perimeter surfaces are called *grain boundaries*. Grains form as the metal solidifies; however, they also may grow or rearrange themselves while the metal solidifies. Grain size is covered in ASTM Specification E 112.
>
>*Hardenability.* This is a characteristic of steel that allows it to be hardened, by some form of heat-treating process.

3.2 Metallic Materials

Metals used for process piping systems can be divided into two groups: ferrous (iron and iron-base alloys) and nonferrous (all other metals and alloys).

The subject of metallurgy covers the extraction of metals from ores and the combining of them with other metals and nonmetals, heat treatment, and processing of metals into useful engineering materials in various product forms, forgings, castings, bars, shapes.

The vast majority of piping materials are made from ferrous metals. Iron is one of the most commonly used metals, but it is rarely found in its purest form, and ferrous metals are defined as those that contain iron as the base metal. The properties of ferrous metals may be changed by adding various alloying elements. The chemical, mechanical, and physical properties need to be combined to produce a metal to serve a specific purpose. The basic ferrous metal form is pig iron, which is produced in a blast furnace charged with an iron ore, coke, and limestone. Iron can be found in the form of various mineral oxides, the principal ones being hematite, limonite, magnetite, and faconite. All ferrous metals are magnetic and give limited resistance to corrosion. The most commonly used ferrous metals are cast iron, mild steel, high-speed steel, high-tensile steel, and stainless steel.

3.3 Alloying of Steel

Carbon steel can be combined, alloyed, with a number of other elements that modify the chemical composition to obtain a wide selection of desired mechanical and physical properties and create engineering materials. The following list identifies the known effects of adding certain elements, in known quantities to steel:

> *Aluminum (Al).* Aluminum is an active deoxidizer used in producing steel. It is used to control inherent grain size.
>
> *Boron (B).* The addition of boron improves hardenability.
>
> *Carbon (C).* An increase in the carbon content of steel alloys usually produces higher ultimate strength and hardness; however, this lowers ductility and toughness. Carbon increases air-hardening tendencies and weld hardness.

Chromium (Cr). Chromium increases steel's response to heat treatment. It also increases depth of hardness penetration. Most chromium-bearing alloys contain 0.50–1.50% chromium. Stainless steels contain chromium in large quantities, 12–25%, frequently in combination with nickel, and possess increased resistance to oxidation and corrosion.

Columbium (Nb). Columbium in 18-8 stainless steel has an effect similar to titanium in making the steel immune to harmful carbide precipitation and the resultant intergranular corrosion.

Copper (Cu). Copper normally is added in amounts of 0.15–0.25% to improve resistance to atmospheric corrosion and increase tensile and yield strength with only a slight loss in ductility. Higher-strength properties can be obtained by precipitation hardening copper-bearing steel.

Iron (Fe). Iron contains other elements in varying quantities that produce the required mechanical properties. Iron lacks strength, but it is very ductile and soft, although it does not respond very well to heat treatment to an appreciable degree.

Lead (Pb). Lead in steel greatly improves its machinability. When the lead is finely divided and uniformly distributed, it has no known effect on the mechanical properties of the steel in the strength levels most commonly specified. It is usually added in amounts of 0.15–0.35%.

Manganese (Mn). Next in importance to carbon is manganese. It normally is present in all steel and functions both as a deoxidizer and to impart strength and responsiveness to heat treatment. Manganese usually is present in quantities from 0.5 to 2%, but certain special steels are made in the range of 10–15%.

Molybdenum (Mo). Molybdenum adds to the penetration of hardness and increases toughness. Molybdenum helps steel to resist softening at high temperatures and is an important means of assuring high creep strength. It generally is used in comparatively small quantities, ranging from 0.10 to 0.40%.

Nickel (Ni). Nickel increases strength and toughness but has little effect on hardenability. It is added in quantities of 1 to 4%, although higher quantities are possible. Steels containing nickel usually have more impact resistance, especially at low temperatures. Certain stainless steels employ nickel up to about 20%.

Phosphorus (Ph). Phosphorus is present in all steel. It increases yield strength and reduces ductility at low temperatures. Phosphorus is believed to increase resistance to atmospheric corrosion.

Silicon (Si). Silicon is one of the common deoxidizers used in the manufacturing of steel. It also may be present in varying quantities up to 1% in finished steel and has a beneficial effect on certain properties, such as tensile strength. It is used in special steels in the range of 1.5 to 2.5% silicon to improve the hardenability. In higher percentages, silicon is added as an alloy to produce certain electrical characteristics in the so-called silicon electrical steels and finds certain applications in some tool steels, where it seems to have a hardening and toughening effect.

Sulfur (S). Sulfur is an important element in steel, because when present in relatively large quantities, it increases machinability. The amount generally used for this purpose is from 0.06 to 0.30%. Sulfur is detrimental to the hot forming properties.

Titanium (Ti). Titanium is added to 18-8 stainless steels to make them immune to harmful carbide precipitation. It sometimes is added to low-carbon sheets to make them more suitable for porcelain enameling.

Tungsten (W). Tungsten is used as an alloying element in tool steel and tends to produce a fine, dense grain, when used in relatively small quantities. In larger quantities of between 17 and 20% and in combination with other alloys, it produces a high-speed steel that retains its hardness at the high temperatures developed in high-speed cutting. It usually is used in combination with chrome or other alloying elements.

Vanadium (V). Vanadium steels have a much finer grain structure than steels of a similar composition without vanadium, which gives additional strength and toughness.

3.4 Types of Steel

There are basically three general groups of carbon steel, based on their carbon content, which can vary from between 0.05 and 1.0 weight percent: low-carbon steels (0.05–0.25% carbon), medium-carbon steels (0.25–0.50% carbon), and high-carbon steels (0.50% and greater carbon content).

3.4.1 Mild (Low-Carbon) Steel

Mild steel is the most commonly used ferrous metal. Its major properties are toughness, high tensile strength, and ductility. It contains approximately 0.15–0.25% carbon. Because of the low carbon content, it cannot be hardened and tempered. It must be case hardened. It is normally used in manufacturing of girders, plates, nuts and bolts, and other general steel products.

3.4.2 Medium-Carbon Steel

Medium-carbon steel has a carbon content of approximately 0.25–0.50%. It is stronger and harder than mild steels but has less ductility, toughness, and malleability. It is used in making steel ropes, wire, garden tools, springs, and the like.

3.4.3 High-Carbon Steel

High-carbon steel is a ferrous metal that contains approximately 0.50% or more carbon. It is the hardest of the carbon steels but is less ductile, tough, and malleable.

3.4.4 High-Tensile Steel

High-tensile steel is a very strong, very tough ferrous metal used exclusively for manufacturing precision steel products, like gears, shafts, and engine parts. This is one of the most frequently used ferrous metals in industry because of its strength, hardness, and toughness.

3.4.5 Stainless Steel

Stainless steels are used for their corrosion-resistance properties or for subzero centigrade temperatures. Stainless steels are those ferrous alloys that contain a minimum of 12% chromium.

Stainless steel types can be divided into the following groups or subcategories: austenitic, martensitic, ferritic, duplex, precipitation hardening, and super alloys.

Austenitic

Austenitic grades are the most commonly used stainless steels for process piping components, although over the last 10 years, the use of duplex is becoming more popular, bearing the total life-cycle cost. Despite the iron present, the austenitic grades are not magnetic. The most common austenitic alloys are iron/chromium/nickel steels, widely known as the 300 series. The austenitic stainless steels, because of their high chromium and nickel content, are the most corrosion resistant of the stainless group, providing unusually fine mechanical properties. They cannot be hardened by heat treatment but can be hardened significantly by cold working.

Straight Grades

The straight grades of austenitic stainless steel contain a maximum of 0.08% carbon. If the stainless steel type meets the physical requirements of straight grade, there is no minimum carbon requirement.

The most commonly used austenitic stainless steel grades for piping material components are 304, 316, and 321.

Low-Carbon Grades

The L suffix after the grade number, for example, 316L, signifies that it is low in carbon. The low carbon content helps with the weldability and provides extra corrosion resistance after welding. The L grades are more expensive; however, the higher carbon content gives the steel greater strength. It is possible to specify dual certified stainless steel, 316/316L. This material has the mechanical strength of 316 and the chemical composition of 316L.

The most commonly used austenitic stainless steel grades for piping material components are 304L, 316L, and 321L.

High-Carbon Grades

The *H* suffix after the grade number, for example, 304H, means that the grade can be used at temperatures higher than that of 304.

Table 3–1 summarizes the austenitic steel grades.

Martensitic

Martensitic grades, also known as 400 series, were developed to provide a stainless steel type that would be corrosion resistant and hardenable by heat treating. The martensitic grades are straight chromium steels containing no nickel. Unlike the 300 series stainless steels, they are magnetic. The martensitic grades are specified where hardness, strength, and wear resistance are required.

Table 3–2 summarizes the grades of martensitic stainless steels.

Table 3–1 Austenitic Steel Grades

Type 304	SS 304 is probably the most internationally used of austenitic grades, although SS 316 is becoming more popular. SS 304 contains approximately 18% chromium and 8% nickel. It is used for chemical processing equipment for the food, dairy, and beverage industries; for heat exchangers; and for use with milder chemicals.
Type 316	SS 316 contains 16–18% chromium and 11–14% nickel. It has molybdenum added to the nickel and chrome of the 304. The molybdenum is used to control pit-type attack. Type 316 is used in chemical processing, pulp and paper processing, food and beverage processing and dispensing, and in more-corrosive environments. The molybdenum must be a minimum of 2%.
Types 321, 347	These stainless steel types have been developed for better corrosive resistance at high temperatures, above 800°F. Type 321 is made by the addition of titanium, and Type 347 is made by the addition of tantalum/columbium. These are the stainless steel types least commonly used in the oil and gas process industries.

Table 3–2 Austenitic Steel Grades

Type 410	The basic martensitic grade, containing the lowest alloy content of the three basic stainless steels (304, 430, and 410), is a low-cost, general-purpose, heat-treatable stainless steel used widely where corrosion is not severe (air, water, some chemicals, and food acids). Typical applications include highly stressed parts needing the combination of strength and corrosion resistance, such as fasteners.
Type 410S	SS 410S contains lower carbon than SS 410 and offers improved weldability but lower hardenability. SS 410S is a general-purpose corrosion- and heat-resisting chromium steel recommended for corrosion-resisting applications.
Type 414	SS 414 has nickel added (2%) for improved corrosion resistance. Typical applications include springs and cutlery.
Type 420	SS 420 contains increased carbon to improve mechanical properties. Typical applications include surgical instruments.
Type 440	SS 440 has further increased chromium and carbon to improve toughness and corrosion resistance. Typical applications include instruments.

Ferritic Grades

Ferritic grades, also 400 series, stainless steel are the least specified in the oil and gas industry. This type of stainless steel also is magnetic but cannot be hardened or strengthened by heat treatment. Generally, these steels are more corrosion resistant than the martensitic grades but inferior to the austenitic grades. Ferritic grades are straight chromium steels with no nickel and most commonly are used for decorative purposes.

Table 3–3 summarizes the grades of ferritic stainless steels.

Duplex Grades

Duplex grades are the newest of the stainless steels, and their use has increased over the last 15 years. This material is a combination of austenitic and ferritic materials. The material has higher strength and superior resistance to stress corrosion cracking than the stainless steels mentioned previously.

Table 3-3 Ferritic Steel Grades

Type 430	SS 430 is the basic ferritic grade, with slightly less corrosion resistance than SS 304.
Type 405	SS 405 has lower chromium and added aluminum to prevent hardening when cooled from high temperatures.
Type 409	SS 409 contains the lowest chromium content of all stainless steels and therefore is the least expensive. It is used in noncritical corrosive environments.
Type 434	SS 434 has molybdenum added, which improves corrosion resistance.
Type 436	SS 436 has columbium added for corrosion and heat resistance.
Type 442	SS 442 has increased chromium to improve scaling resistance.
Type 446	SS 446 contains even more chromium to further improve corrosion and scaling resistance at high temperatures.

Initially, duplex stainless steels were introduced for offshore applications; however, their use has become more widespread, as their advantages have been witnessed in service. The benefit from the use of duplex stainless steel is that it combines the basic toughness of the more common austenitic stainless steels with the higher strength and improved corrosion resistance of ferritic steels.

A significant characteristic of duplex stainless steel is that its pitting and crevice corrosion resistance is superior to that of standard austenitic alloys.

The two most commonly specified duplex stainless steels are duplex stainless steel and super duplex steel.

Precipitation Hardening Grades

Precipitation hardening grades of stainless steel offer a unique combination of fabricatability, strength, acceptance of heat treatment, and corrosion resistance.

The most commonly used grades are 17-4PH, which contains 17% chromium and 4% nickel, and 15Cr-5Ni (15-5PH). The martensitic precipitation-hardenable stainless steels are used to form products like bars, rods, wire, and forgings.

Super Stainless Steels

Super stainless steels are used when austenitic and ferritic/austenitic stainless steels are inadequate to withstand corrosion attack. They contain very large percentages of nickel or chrome and molybdenum. This makes them much more expensive than the usual 300 series alloys and not as readily available. These alloys include alloy 20 and Hastelloy.

Alloy steels are considered to be steels to which one or more alloying elements, other than carbon, have been added to give them mechanical and chemical properties that are different than those of carbon steels.

Steel is considered an alloy steel when the levels of manganese, silicon, or copper exceed the maximum limits for the carbon steels or when minimum quantities of other alloying elements, such as chromium, molybdenum, nickel, copper, cobalt, niobium, vanadium, or others, are added intentionally.

The higher levels of alloying in steels used in the piping industry are ferritic and martensitic stainless steels. These are steels alloyed with chromium contents above about 12%. The presence of chromium gives these steels good corrosion resistance, which increases with the addition of more chromium.

Austenitic stainless steels are the ones most commonly used in the process industry, and they possess an excellent combination of strength, ductility, and corrosion resistance.

These materials are referred to as *precipitation hardenable stainless steels*. Both martensitic and austenitic stainless steels can be enhanced in this manner. As annealed, these materials are soft and readily formed. When fully hardened through aging heat treatments, they attain their full strength potential.

3.5 Steel Heat-Treating Methods

A number of heat treatment methods can be used to modify specific properties of steel, such as the hardness and ductility, improvement of machinability, stress relief, or obtaining high strength and impact properties.

The heat treatments of steel most commonly employed are annealing, normalizing, spheroidizing, and hardening (quenching) and tempering.

3.5.1 Annealing

The annealing of a metal has one of the following objectives:

1. To soften the steel and to improve machinability.

2. To relieve internal stresses induced by some previous treatment (rolling, forging, uneven cooling).

3. To remove coarseness of grain.

Several types of annealing processes are used on carbon and low-alloy steel. These are generally referred to as *full annealing, process annealing,* and *spheroidizing annealing.*

Annealing consists of heating a metal to a suitable temperature and holding it there, followed by cooling at a suitable rate. It is used to soften but also to produce desired changes in other properties in microstructure.

Annealing improves machinability, facilitation in cold working, mechanical or electrical properties, and dimensional stability.

The time/temperature cycles used vary widely both in the maximum temperature to be attained and the cooling rate employed, depending on the composition of the metal, its condition, and the results desired.

When the term is used without qualification, full annealing is implied. When applied only for the relief of stress, the process is properly called *stress relief.*

In full annealing, the steel is heated to just above the upper critical temperature, held for a sufficient length of time to fully austenitize the material structure, then allowed to cool at a slow, controlled rate in the furnace. A full annealing provides a relatively soft, ductile material free of internal stresses.

Process annealing, sometimes referred to as *stress relief*, is carried out at temperatures below the lower critical temperature. This treatment is used to improve the ductility and decrease residual stresses in work-hardened steel.

Spheroidizing softens the steel and improves its machinability.

3.5.2 Normalizing

The normalizing heat treatment cycle involves heating slowly to the normalizing temperature for that particular steel, holding it at a temperature sufficient to allow homogenization, then air cooling to room temperature. Normalizing relieves the internal stresses caused by previous cold working and produces softness and ductility. The steel is left harder and with higher tensile strength than after annealing.

3.5.3 Hardening

Steel is hardened by a rapid cool down (quenching) from within or above the critical temperature range of the metal. The temperatures are the same as those given for full annealing.

3.5.4 Tempering

Tempering is classified as a secondary heat treatment performed on some normalized and almost all hardened steels. The object of tempering is to remove some of the brittleness by allowing certain solid-state transformations to occur. The steel is heated to a predetermined temperature, which is always below the lower critical temperature. This is followed by a controlled rate of cooling. In most cases, tempering reduces the hardness of the steel, increases its toughness, and eliminates residual stresses.

3.6 Nonferrous Metals in Alloying

Commonly used nonferrous pure metals employed in alloying include aluminum, brass, copper, tin, titanium, and zinc and their

Table 3–4 Nonferrous Metals Used in Alloying Steel

Name	Properties
Aluminum	Grayish white, soft, malleable, conductive to heat and electricity, it has a degree of corrosion resistance.
Brass	Yellow in color, it tarnishes very easily; Harder than copper.
Copper	Red, tough, ductile, high electrical conductor, corrosion resistant, it can be worked hard or cold.
Lead	The heaviest common metal, it is soft, malleable, bright and shiny when new but quickly oxidizes to a dull gray; resistant to corrosion.
Tin	White and soft, corrosion resistant.
Titanium	Light, strong, corrosion resistant with a white-silvery metallic color, it is used in strong lightweight alloys, especially iron and aluminum.
Zinc	A layer of oxide protects it from corrosion, it is bluish-white, easily worked.

various alloyed types. Precious metals such as silver, gold, and platinum are also nonferrous, but their cost makes them prohibitively expensive. Table 3–4 describes some of the nonferrous metals used in alloying steel.

3.7 Material Specifications

There are numerous material standards for metallic piping components, but the two most commonly used in ASME B31 code projects are from the American Society for Testing and Materials (ASTM) and the Unified Numbering System (UNS).

3.7.1 American Society for Testing and Materials

The scope of the ASTM body is discussed in Chapter 1, complete with a list of the extent of the coverage of the 67 volumes. Many of these are not relevant to process piping systems and the process industry.

The most commonly referred to volumes for process piping systems designed to ASME B31 codes are the following:

Section 1, Iron and Steel Products.

> Volume 01.01, Steel—Piping, Tubing, Fittings.
>
> Volume 01.02, Ferrous Castings; Ferroalloys.
>
> Volume 01.03, Steel—Plate, Sheet, Strip, Wire.
>
> Volume 01.04, Steel—Structural, Reinforcing, Pressure Vessel, Railway.
>
> Volume 01.05, Steel—Bars, Forgings, Bearing, Chain, Springs.
>
> Volume 01.06, Coated Steel Products.
>
> Volume 01.07, Shipbuilding.

Section 2, Nonferrous Metal Products.

> Volume 02.01, Copper and Copper Alloys.
>
> Volume 02.02, Aluminum and Magnesium Alloys.
>
> Volume 02.03, Electrical Conductors.
>
> Volume 02.04, Nonferrous Metals—Nickel, Cobalt, Lead, Tin, Zinc, Cadmium, Precious, Reactive, Refractory, Metals, and Alloys.
>
> Volume 02.05, Metallic and Inorganic Coatings; Metal Powders, Sintered P/M Structural Parts.

Section 3, Metals Test Methods and Analytical Procedures.

> Volume 03.01, Metals—Mechanical Testing: Elevated and Low-Temperature Tests, Metallography.
>
> Volume 03.02, Wear and Erosion, Metal Corrosion.
>
> Volume 03.03, Nondestructive Testing.

Volume 03.04, Magnetic Properties; Metallic Materials for Thermostats, Electrical Heating and Resistance, Heating, Contacts, and Connectors.

Volume 03.05, Analytical Chemistry of Metals, Ores, and Related Materials (I).

Volume 03.06, Analytical Chemistry of Metals, Ores, and Related Materials (II).

Section 8, Plastics.

Volume 08.01, Plastics (I): C 177–D 1600.

Volume 08.02, Plastics (II): D 1601–D 3099.

Volume 08.03, Plastics (III): D 3100–Latest.

Volume 08.04, Plastic Pipe and Building Products.

Section 9, Rubber.

Volume 09.01, Rubber, Natural, and Synthetic—General Test Methods; Carbon Black.

Volume 09.02, Rubber Products, Industrial—Specifications and Related Test Methods; Gaskets; Tires.

Section 00, Index.

Volume 00.01, Subject Index and Alphanumeric List.

3.7.2 Unified Numbering System of Ferrous Metals and Alloys

The Unified Numbering System for Metals and Alloys provides methods of correlating many internationally used metal and alloy numbering systems currently published by engineering bodies, societies, trade associations, and producers of metals and alloys. This numbering system is not limited to the oil and gas industry; it is commonly referred to through most engineering industries, especially in the United States and Europe.

The system helps avoid the confusion caused by the use of more than one identification number for the same type of metal or alloy. Such uniformity provides an efficient method for referencing and cross-referencing material types.

The UNS identifies nine series of designations for ferrous metals and their alloys. Each UNS designation consists of a single-letter prefix followed by five digits. In most cases the letter is suggestive of the family of metals identified; for example, *A* for aluminum, *F* for cast irons, *T* for tool steel, and *S* for stainless steels.

The cross-referenced specifications in Table 3–5 are representative only and are not necessarily a complete list of specifications applicable to a particular UNS designation. The table is an outline of the organization of UNS designations.

Table 3–5 UNS Series

UNS Series	Metal
A00001 to A99999	Aluminum and aluminum alloys
C00001 to C99999	Copper and copper alloys
D00001 to D99999	Specified mechanical property steels
E00001 to E99999	Rare earth and rare earthlike metals and alloys
F00001 to F99999	Cast irons
G00001 to G99999	AISI and SAE carbon and alloy steels (except tool steels)
H00001 to H99999	AISI and SAE H-steels
J00001 to J99999	Cast steels (except tool steels)
K00001 to K99999	Miscellaneous steels and ferrous alloys
L00001 to L99999	Low-melting metals and alloys
M00001 to M99999	Miscellaneous nonferrous metals and alloys
N00001 to N99999	Nickel and nickel alloys
P00001 to P99999	Precious metals and alloys
R00001 to R99999	Reactive and refractory metals and alloys
S00001 to S99999	Heat and corrosion resistant (stainless) steels
T00001 to T99999	Tool steels, wrought and cast
W00001 to W99999	Welding filler metals
Z00001 to Z99999	Zinc and zinc alloys

CHAPTER 4

Roles and Responsibilities

For an engineering, procurement, and construction (EPC) contract to commence, a multidiscipline organization is created that comprises personnel from the following departments: Process, Civil, Structural, Piping, Mechanical, Instrumentation, Electrical, and Project Services (Costing and Planning).

The basic organizational framework for the piping group has changed very little since the advent of computer-aided design. Piping designers now create their piping systems digitally instead of as manually prepared drawings. The engineering content, however, remains the same as it was in the past. There is one significant new role in the Piping Department: the piping CAD coordinator.

Of these various groups, the Piping Department usually is the largest and generally comprises four principal sections, each reporting to a project lead piping engineer (PEL):

- Piping materials engineering (PME) group.
- Piping design group (PDG).
- Piping materials control (PMC).
- Piping stress engineering (PSE).

The project lead piping engineer, in turn, reports to the project engineering manager, who is responsible for all engineering activities and all disciplines on the project: process, civil, structural, mechanical, and instrumentation.

No two projects have exactly the same organizational matrix: however, there always are similarities. The chapter is divided into the following sections:

> 4.1. Lead Piping Engineer.
>
> 4.2. Piping Materials Engineering Group.
>
> 4.3. Piping Design Group.
>
> 4.4. Piping Materials Control Group.
>
> 4.5. Piping Stress Engineering Group.
>
> 4.6. Other Engineering Disciplines Involved.

4.1 The Lead Piping Engineer

The project lead piping engineer is a manager of technical personnel who has a good general background on the subject of piping. It is essential that this person be a good manager of personnel to get the best out of his or her team. This is a technical supervisory role, because this individual is responsible for the complete project piping group and must ensure that the delegated work is carried out by competent individuals in a timely manner.

The responsibilities of a lead piping engineer are to

- Assemble and supervise a piping group capable of executing the project.
- Create a project execution plan for the piping group.
- Create a project piping engineering organization chart.
- Delegate responsibility to the piping design, piping materials, engineering, piping stress, and the piping materials control groups.

- Estimate the work hours of piping engineering needed to complete the project, review this with engineering management, and maintain this schedule throughout the project.
- Develop and maintain the piping engineering work plan.
- Supervise the creation of the piping material specifications and piping standards.
- Define the document distribution matrix within the project.
- Attend kickoff meetings, project staff, client, construction, flow diagram review, and other meetings as required.
- Act as the focal contact point with the client and other discipline lead engineers on all piping issues.
- Be responsible for the implementation of the project procedure manual within the piping group.
- Maintain design deviations in the job scope and estimate additional hours spent on these activities.
- Maintain project budget control and planning regarding labor, trends, and change orders.
- Monitor, control, schedule, and report pipe shop fabrication and delivery (when applicable).
- Periodically report piping progress to project and discipline management.
- Issue all piping design deliverables, such as specifications, drawings, standards, and requisitions.
- Coordinate, resolve, or delegate on-site queries.
- Create a piping engineering project completion report.
- Assemble a lessons learned report for the project.

4.2 Piping Materials Engineering Group

Depending on the size of the project, this is either a piping materials engineer or, on larger projects, a team of engineers responsible for the integrity of the materials of construction used in a piping system. The scope of this person or group covers the following areas.

4.2.1 Project Lead Piping Materials Engineer

The lead piping material engineer reports directly to the lead piping engineer. The responsibilities of a lead piping materials engineering are to

- Supervise the personnel within the piping material engineering group.
- Estimate the work hours plus planning and staffing levels for the group.
- Develop piping material specifications, fabrication, and testing procedures.
- Act as the focal point for communication with other piping groups and groups from other disciplines.
- Act as the focal point for communication with the Procurement Department.
- Maintain deviations in the scope of supply from vendors.
- Attend material engineering in meetings with suppliers, subcontractors, and the client.
- Submit material engineering specifications and line list to the lead project piping engineer for issuing.
- Review and approve new manufacturers of piping material.
- Review and approve the bidders list.
- Develop and submit recommendations for materials inspection, identification, certification, and expediting to the project piping engineering leader.
- Compile and issue a lessons learned report at the end of the project.

The lead piping materials engineer usually starts on the project very early, sometimes even before it is awarded and remains through to the end. During peak times, this engineer is assisted by other PMEs.

4.2.2 Senior Piping Materials Engineer

The senior piping materials engineer reports directly to the lead piping material engineer. The responsibilities of a piping materials engineer are to

- Develop piping material, insulation, and coatings specifications.
- Develop project piping classes.
- Determine cleaning requirements for certain services.
- Develop and maintain the line list.
- Create valve data sheets.
- Create data sheets for piping specialty items.
- Develop and maintain a piping specialty list for out-of-spec piping components.
- Calculate pipe wall thicknesses.
- Assist in the compilation of the commodity catalogue.
- Prepare requisitions packages for piping components.
- Review technical bid evaluations for acceptability.
- Agree on concession requests with vendors.

The deliverables of the piping material engineering group are

- Project piping classes.
- Project piping specifications: fabrication, testing, painting.
- Valve data sheets for manual valves and on/off valves.
- Technical requisition packages for all piping components.
- Technical bid evaluations.
- Concession requests.
- Deviation requests.

4.3 Piping Design Group

Generally the largest group, it is made up of piping supervisors, sometimes called *squad bosses*, and piping designer-checkers, who are responsible for pipe modeling, and experienced checkers. Depending on the size of the project there could be a lead piping layout engineer, piping coordinator, lead designers for each area or process unit, and underneath them, piping designers and checkers.

The group is responsible for the piping layouts within the plant. The plant is divided into process areas and further divided into units. Generally, this group is lead by one individual, the lead piping design supervisor, who reports directly to the lead piping engineer.

The responsibilities of the piping design supervisor are to

- Supervise assigned piping design personnel.
- Develop piping design work-hour estimates plus planning and staffing levels.
- Develop piping design administrative procedures.
- Develop piping design technical standards and procedures.
- Assist in the development and maintenance of plot plans.
- Report to the PEL with progress of the piping design group.
- Act as the focal interface point with other piping groups.
- Act as the focal interface point with those from other engineering disciplines.
- Attend CAD model reviews if required.
- Attend meetings on hazardous operations if required.
- Maintain and distribute piping job notes.
- Approve drawings generated from the piping design group.
- Prepare a lessons learned report for the piping design group.

4.3.1 Project Piping Area/Unit Supervisor (Squad Boss)

The piping area/unit supervisor reports directly to the lead piping design supervisor. The piping area/unit supervisor's job is limited to specific areas or units on the project. The responsibilities of the piping area/unit supervisor (squad boss) are to

- Supervise assigned piping design personnel.
- Develop piping design work-hour estimates plus planning and staffing levels.
- Implement piping design administrative procedures.
- Implement piping design technical standards and procedures.
- Assist in the development and maintenance of plot plans.
- Report to the PEL with progress of the piping design group.
- Act as the focal interface point with other piping groups.
- Act as the focal interface point with those from other engineering disciplines.
- Attend CAD model reviews if required.
- Attend meetings on hazardous operations if required.
- Distribute piping job notes.
- Check drawings generated by the piping design group.
- Prepare a lessons learned report for the piping design group.

4.3.2 Project Piping CAD Coordinator

The piping CAD coordinator reports directly to the lead piping engineer. The piping CAD coordinator's job covers all areas and units on the project. This is a relatively new role in the piping group. It is an essential position since the introduction of CAD systems. This coordinating role ideally should be filled by someone with a strong technical background and familiar with the latest CAD information technology.

The responsibilities of the piping CAD coordinator are to

- Determine requirements for the sequence of data production and sharing of databases.
- Distribute approved piping CAD symbols.
- Create and maintain the catalogue of piping components.
- Meet with the corporate CAD Department on piping-related hardware and software problems and requests.
- Provide CAD support to piping designers.
- Prepare or gather and distribute approved CAD-related written materials, manuals, procedures, and CAD notes to assigned designers.
- Provide a central focus for CAD within the project piping group.
- Assist the company CAD Department with archiving project piping CAD files.
- Assist the piping design group with scheduling, determining workstation requirements, and forecasting CAD-related costs and estimates.

4.3.3 Project Piping Designers-Checkers

The piping designer-checkers report directly to the project piping design supervisor or piping area/unit supervisor. The piping designer's job is limited to specific areas or units on the project. This designer must be familiar with project specifications, administrative and technical practices, design instructions, plot plans, flow diagrams, and supplier information.

The responsibilities of piping designer-checkers are to

- Complete CAD layouts using project specifications, standards, procedures, plot plans, piping and instrument diagrams (P&IDs), line list, stress information, and vendor data.

- Meet with those from other engineering disciplines.
- Extract computer drawings and isometrics (see Figure 4–1).
- Attend CAD model reviews if required.
- Attend meetings on hazardous operations if required.
- Check, back-check, and revise computer drawings and isometrics.
- Check vendor equipment drawings.
- Make the supervisor aware of lessons learned.

The piping design group deliverables are

- Project plot plan.
- Equipment layout drawings.
- Manual piping study drawings.
- Piping sections.
- 2D and 3D piping CAD models.
- Isometric extraction of the model.
- Piping standard assemblies.
- Piping tie-in drawings.
- A piping drawing index.
- Heat/steam tracing drawings.
- A piping general requirements specification.

4.4 Piping Materials Control Group

This group of designers works in conjunction with the piping materials engineering group and the piping design group to create bills of materials, which accurately specify and quantify the piping materials of construction, based on the latest pipe routing.

146 Chapter 4—Roles and Responsibilities

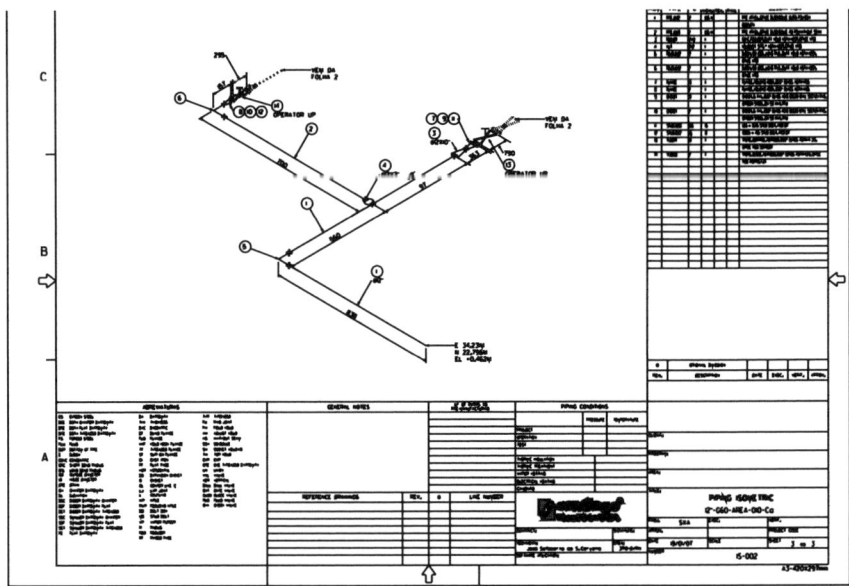

Figure 4–1 *An example of an isometric extracted from a 3D CAD model. (Printed with the permission of Bentley Systems Incorporated.)*

During design development, material takeoffs are made, and with these quantities, requisitions are created and issued to the Procurement Department, which contacts suitable manufacturers for bids.

4.4.1 Project Lead Piping Materials Controller

The lead piping material controller reports directly to the lead piping engineer. The lead piping material controller's job covers all units.

The responsibilities of the lead piping material controller are to

- Supervise the piping material controllers on the project.
- Develop piping design work-hour estimates plus planning and staffing levels.
- Issue material control reports.
- Issue reports for quantities for requisitions.

- Communicate with other piping groups and those from other engineering disciplines.
- Prepare and maintain the required on site (ROS) date schedule.

4.4.2 Project Piping Materials Controller

The project piping material controller reports directly to the lead piping material controller. The piping material controller's job covers all units.

The responsibilities of the piping material controller are to

- Maintain material takeoffs during the various stages of the project—bulk, intermediate, and final.
- Carry out bulk valve counts from P&IDs.
- Release quantities for the various requisitions and subsequent "top-ups."
- Maintain material received records.

The piping material control group deliverables are

- Various piping material takeoffs (MTOs).
- Various piping bills of materials (BOMs).
- Piping material summaries by piping class.
- Manual valve count.
- Piping requisition quantities.
- Piping purchase order quantities.
- Required on site dates.

4.5 Piping Stress Engineering Group

4.5.1 Project Lead Piping Stress Engineer

The lead piping stress engineer reports directly to the lead piping engineer. The lead stress engineer's job covers all units.

The responsibilities of the lead stress engineer are to

- Develop the piping stress engineering philosophy and specifications.
- Supervise piping stress personnel on the project.
- Develop and maintain the critical line list that determines the scope of stress analysis on the project.
- Develop piping design work-hour estimates plus planning and staffing levels.
- Report to PEL on all stress issues and progress.
- Act as the focal point for stress issues to other piping groups and those from other engineering disciplines.
- Approve and issue all stress documents, calculations, and sketches.
- Approve quantities and technology for all stress-related supports and equipment.
- Act as the focal point for all site queries related to stress.
- Prepare and maintain a lessons learned report.

4.5.2 Project Piping Stress Engineer

The piping stress engineer reports directly to the lead piping stress engineer. The piping stress engineer's job covers specific areas or units.

The responsibilities of the piping stress engineer are to

- Create and maintain piping flexibility files for critical lines.
- Create technical specifications for expansion joints, slide plates, spring hangers, shock arrestors, and other stress-related equipment.
- Review requests for quotation and purchase requests for cost and technical completeness for stress-related components.
- Resolve site queries regarding stress.

The piping stress engineering group is responsible for the stress analysis of the piping systems considered to be critical. This criticality is determined by a set of "rules" that takes into consideration

- Pipe size.
- Pressure.
- Temperature.
- Service.
- Location.
- Seismic conditions.

Piping systems that fall into the critical category are analyzed to ensure that these lines are not overstressed. The calculations are carried out either manually or using computer software like Caesar. Lines that are overstressed have to be either rerouted, anchored, or guided. In certain cases, expansion joints have to be used.

Pipe supporting is another specialist activity under the control of the piping stress engineering group or the piping design group. Pipe support specialists work closely with the PSE and the piping design group, and they are responsible for the selection of the type and the location of pipe supports. These pipe supports are fabricated from steel, or more complex supports are of a proprietary design.

The piping stress group deliverables are

- Stress calculation (manual and computed).
- Stress packs.
- Calculation of nozzle loads.
- Spring hanger specifications and quantities.
- Expansion joint specifications.
- Slide plate specifications.
- Technical bid evaluation for stress-related equipment.

4.6 Other Engineering Disciplines Involved

The design of a process plant is a team effort, and although the piping discipline is very important, it requires equal effort from several other engineering and nonengineering departments. A typical project would comprise the following engineering disciplines:

- Architectural buildings.
- Civil.
- Cost.
- Electrical.
- Estimating.
- Instrument.
- Mechanical.
- Piping.
- Planning.
- Process.
- Safety (fire protection).
- Structural.
- Vessels.

Of these disciplines, the Piping Department usually is the largest, followed by Structural and Instrumentation Departments. As an integrated engineering team, interface among disciplines is very important in allowing the project to progress to mechanical completion.

Piping is a big subject, and of all the disciplines, it probably has the most interfaces: however, the groups most commonly closely worked with during the design phase are civil, instrument, mechanical, process, and structural engineering. Transfer of current information is bidirectional, and it is essential that a good working relationship is established among all the disciplines on the project.

Discussed next are examples of the type of information that the piping group requires from each discipline.

4.6.1 Process Engineering

Of all of the major disciplines on the project, the process engineering group is the one from which the piping group takes the most guidance. A process engineering group usually is divided into the following roles and responsibilities. This is one option for the structure of a process engineering group, but there are several alternatives. The group is headed by a project lead process engineer. The primary and secondary processes have

- Lead process engineer—process (all areas and units).
- Lead process engineer—process (specific area or unit).
- Process engineer—process (specific area or unit).

For utilities, there are

- Lead process engineer—utilities (all areas and units).
- Lead process engineer—utilities (specific area, unit, or system).
- Process engineer—process (specific area, unit, or system).

For off-site areas and units, there are

- Lead process engineer—utilities (all system).
- Lead process engineer—utilities (specific system).
- Process engineer—process (specific system).

The process group is responsible for interpreting a client's requirements and creating a series of specifications, drawings, and support calculations that become more defined as the project develops.

The process engineering group has important input into the creation of the materials selection report in conjunction with the project metallurgist. This document defines the selection of the materials of construction for the various piping systems and the pressure and temperature limitations of the materials.

The process group's deliverables that are of importance to the piping group include

- Process flow diagrams.
- Process engineering diagrams.
- Piping and instrument diagrams.
- Process data sheets for equipment.
- Process line list.

Process Flow Diagram

First, basic process flow diagrams (PFDs) are created for each area of the primary process system and the various types of major equipment necessary to achieve the end product. This includes process equipment necessary to alter the state of the process flow, that is, change its temperature, and to move it through the process chain using pumps and compressors.

Process Engineering Diagrams

At an early stage, the process engineering group uses basic process engineering diagrams (PEDs) to create a process line list, process data sheets for equipment, and a piping and instruments diagram.

The P&IDs are fundamental drawings for the piping group, and it is essential that the piping engineers and designers always work to the latest revision. These drawings are schematics without dimensions, so what appears to be a small change on a P&ID can have a dramatic affect on piping design and layout. (See Figure 4–2.)

Information from all disciplines is important, and in many cases, if it is not available, work can be put on hold; but without the P&IDs (see Figure 4–3), the piping group cannot commence detailed design work.

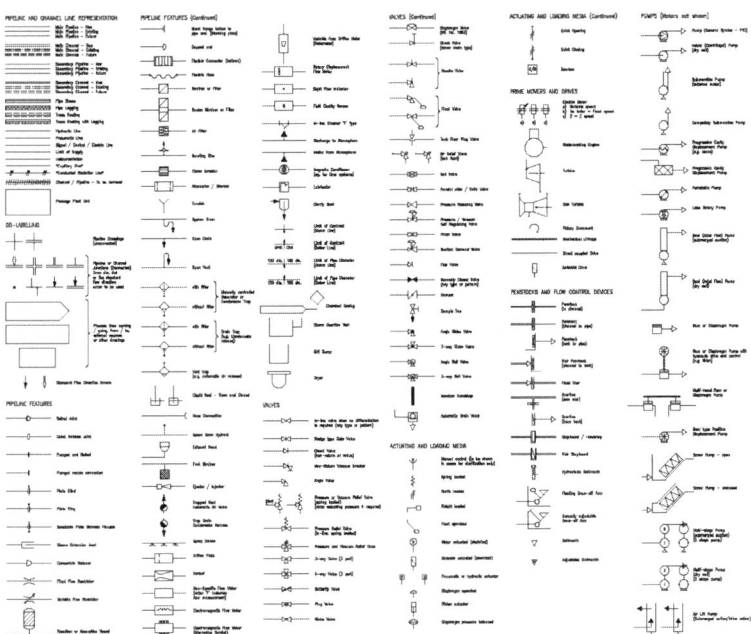

Figure 4–2 *An example of a P&ID legend sheet. (Printed with the permission of Bentley Systems Incorporated.)*

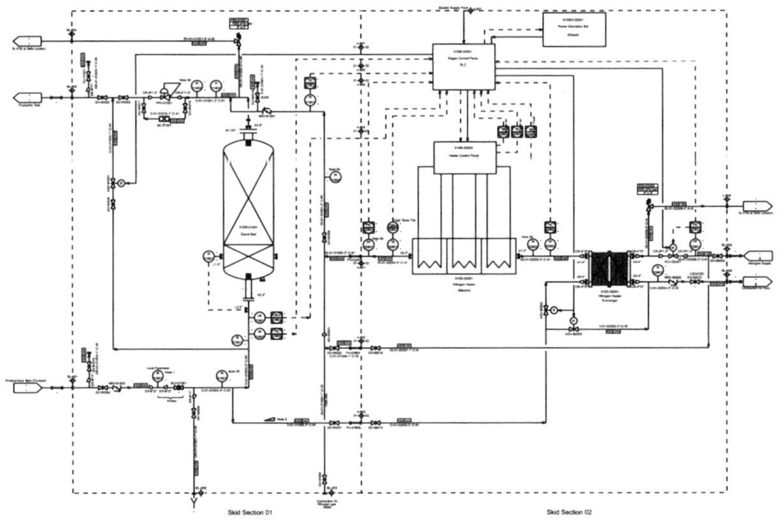

Figure 4–3 *An example of a P&ID. (Printed with the permission of Bentley Systems Incorporated.)*

4.6.2 Mechanical Engineering

The mechanical engineering group basically can be broken down into two sections: rotating equipment (pumps, turbines, and compressors) and static equipment (vessels, columns, tanks, and heat exchangers). These are handled by two distinctly different groups of mechanical engineers with specialist skills, who are responsible for the various types of process equipment. Detailed equipment specifications are developed by this group, based on the process data supplied by the process engineering group, which include capacities, throughput, and design and operating pressures and temperatures.

The piping engineering group has to coordinate its work very closely with the mechanical engineering group. The mechanical engineering group receives detailed dimensional drawings from vendors, and the size and the orientation of the numerous nozzles are identified. This allows the piping designer to pipe up to the correct location and route the pipe accurately.

Within this group are mechanical package engineers, who are responsible for packaged units, which could include several items of equipment, plus the interconnecting pipework.

The mechanical engineering group deliverables that are of importance to the piping group are

- Mechanical data sheets for equipment.
- Vessel data sheets for equipment.
- Vendor dimensional details of all equipment.
- Acceptable nozzle loads.

4.6.3 Instrument Engineering

Instrument engineering covers the control and the measurement of a process system, and that covers the process's flow, pressure, and temperature. The instrument engineering group's deliverables that are of importance to the piping group are instrument hookup details, control valve data sheets, and on/off valve data sheets.

4.6.4 Civil Engineering

For onshore projects, the civil engineering group is involved in numerous interfaces; however, for offshore projects, civil engineering has very little presence. The civil engineering group deliverables that are of importance to the piping design group are

- Topographical drawings.
- Site surveys.
- Foundations plans.
- Roads and paved areas.

4.6.5 Structural Engineering

Close coordination between the Piping and the Structural Departments is essential for both onshore and offshore process projects. The Structural Department comprises structural engineers and designers who are responsible for the structural integrity of the numerous structures, pipe racks, and pipeways required for the mechanical equipment and the interconnecting piping systems. The structural engineering group deliverables that are of importance to the piping design group are structural plans, sections, and details for all areas.

CHAPTER 5

Projects

This chapter discusses project types and project phases. I keep these subjects as simple as possible, because no two projects are the same, but all have a beginning, a middle, and an end. I will avoid discussing in depth numerous possibilities that can happen during a project.

A multitude of books have been written on the project process and project execution; they should be referred to for more detailed information on the two subjects. My suggestion is that, if these subjects are of interest to you, then read as many as you can, because there are numerous options and opinions.

Numerous types of projects require process piping systems, and I concern myself with the ones that a piping design engineer comes across most frequently. The purpose of this chapter is to give an indication of types of projects and their various stages.

5.1 Project Types

A project that requires process piping systems can be divided into two types of construction, either grass roots or Brownfield.

A grassroots project is one where there has been no previous development, or if there are buildings of existing facilities, they are completely flattened and the site is totally cleared. A grassroots project is new, totally fresh, and on a dedicated plot for the facility. An expansion on a new plot that ties into an existing facility also can be classified as a grassroots project.

A Brownfield project, sometimes referred to as a *revamp*, involves modifications to an existing plant to increase its capacity, make it operate more efficiently, or change the refined product.

A hydrocarbon project produces a plant to handle one of the following processes:

- Hydrocarbon refining.
- Oil and gas separation.
- Ammonia production.
- Aromatics production.
- Benzene production.
- Butadiene production.
- Crude oil refining (see Figure 5–1).
- Ethylene production.
- Fertilizer manufacturing.
- Changing gas to liquids (GTL).
- Gasoline blending.
- Liquefied natural gas (LNG) production.
- Paint manufacture.
- Polyethylene production.
- Sulfur recovery.

The feedstock for all of these is hydrocarbons, in either a liquefied, gaseous, or combined state. This is not intended to be a definitive list but one that gives the reader an idea of the diversity of the projects that use hydrocarbons as a feedstock.

There also are many nonhydrocarbon processes:

5.1 Project Types 159

Figure 5-1 *A refinery designed to ASME B31.3. (Printed by permission of Bentley Systems Incorporated.)*

- Water treatment.
- Waste treatment.
- Sewage treatment.
- Pulp paper manufacture.
- Air separation (oxygen, nitrogen, helium, argon).
- Chlorine production.
- Beer distilling.
- Food processing.

Both the hydrocarbon and nonhydrocarbon projects mentioned require pressurized process piping systems that have to be designed, fabricated, erected, inspected, and tested to recognized international codes and associated standards, specifications, and procedures. ASME B31.3 is one such standard, and it can be applied to all of the projects mentioned previously.

5.2 Project Phases

Projects evolve over several years, and they all follow a similar path. The conception to commissioning of mega projects can span two decades. A project is initiated by an operator or end user who recognizes that it has the feedstock, a market for the refined product, and the funding for the project. Local and government approvals must be in place.

A project takes several phases, regardless of the size and the complexity of the process:

- Feasibility.
- Conception.
- Front-end engineering development.
- Detailed engineering.
- Construction.
- Commissioning.
- Startup and handover to the owner.

5.2.1 Feasibility Phase

The feasibility phase determines the technical and commercial viability of the project. This covers availability of the feedstock, a well-defined market, local and government approval, and licenses to construct. A project cannot proceed until all theses points have been covered. In addition, environmental issues have to be addressed. This activity usually is carried out by the operator/end user, but in certain cases it could involve specialist contractors.

5.2.2 Conception Phase

The conception phase determines the basic engineering scope, economics, and the preliminary schedule for the proposed project. This activity usually is carried by out by a specialist engineering contractor, who may follow this project through to startup.

For the piping group, the major deliverables during this phase are as follows. The piping material engineering group prepares a preliminary list of piping classes and preliminary technical piping specifications. The piping design group prepares preliminary piping design specifications and piping studies.

5.2.3 Front-End Engineering Development Phase

The front-end engineering development (FEED) phase develops the basic engineering scope and allows vendors of the major items of equipment to be consulted further so that price and deliveries can be more accurately determined. At the FEED stage, purchase orders can be placed for major items of equipment with deliveries in excess of one year, like large pumps, compressors, and vessels. This activity could be an extension of the conceptual study and carried out by the engineering contractor that worked during the conception phase.

At this stage, piping drawings and documents are issued for approval. For the piping group, the major deliverables during this phase are as follows. The piping material engineering group prepares

- List of piping classes.
- Wall thickness calculations.
- Valve data sheets.
- Piping special data sheets.
- Piping fabrication, installation, inspection and testing specifications.
- Painting specifications.
- Technical specifications for the purchase of all piping components.

and issues concession requests. The piping design group prepares

- General piping design specifications.
- Piping standard assemblies.

- Piping drawing index.
- Piping layout studies.

The piping stress engineering group prepares

- Preliminary piping flexibility analysis for critical lines.
- Preliminary technical specifications for expansion joints, spring supports, slide plates.
- Preliminary data sheets for expansion joints, spring supports, slide plates.

The piping material control group prepares preliminary material take-offs for requisitions.

5.2.4 Detailed Engineering Phase

The detailed engineering phase takes the FEED contract documents as a basis for design and the project is fine-tuned and fully determined with the minimum amount of deviation. At this stage, the remaining purchase orders are placed. The contract for this activity could be one of the following: engineering and procurement (E&P) or engineering, procurement, and construction (EPC).

There are other variations, but these are the two most common. They can be awarded to one company or, what is more common on larger projects, a consortium of specialist engineering and construction companies. The latter spreads the responsibilities and risks of the project to a consortium of contractors.

The EPC contract includes the production engineering phase, during which the project drawings, design specifications, and procedures are developed, approved, and issued for construction (IFC).

For the piping group, the major deliverables during this phase are as follows. The piping material engineering group issues

- Piping classes (IFC).
- Wall thickness calculations.

- Valve data sheets (IFC).
- Piping special data sheets (IFC).
- Piping fabrication, installation, inspection, and testing specifications (IFC).
- Painting specifications (IFC).
- Technical specifications for the purchase of all piping components. (IFC).
- Technical content for piping requisitions (IFC).
- Technical bid evaluations for piping components.

as well as reviews and approves concession requests. The piping design group issues

- General piping design specifications (IFC).
- Piping standard assemblies (IFC).
- Piping drawing index (IFC).
- Piping layout studies (IFC).
- A piping CAD 3D model (see Figure 5–2).
- Piping isometrics (IFC).
- Tie-in list (IFC).
- Tie-in drawings (IFC).

The piping stress engineering group issues

- Piping flexibility analysis for critical lines (IFC).
- Stress isometrics (IFC).
- Technical specifications for expansion joints, spring supports, slide plates (IFC).

Figure 5–2 *A 3D CAD model of a vessel, showing foundation, inlet, outlet process piping, utility pipe work, and platforms. (Printed with the permission of Bentley Systems Incorporated.)*

- Data sheets for expansion joints, spring supports, slide plates (IFC).
- Technical bid evaluations for piping stress components (IFC).

as well as assembles stress packs (IFC) and reviews and approves concession requests regarding piping stress issues. The piping material control group issues

- Intermediate material takeoffs for requisitions (IFC).
- Final material takeoffs for requisitions (IFC).
- The control reports of piping material to the site (IFC).

The procurement phase, covers purchasing, inspection, and expediting. It can be spread over the conception, FEED, and the detailed engineering phases of the project; however, most of the purchase orders for the piping group are placed during the detailed engineering phase. Most piping components are considered to be bulk commodity items, and they cannot be purchased until quantities can be accurately determined, which does not occur until the detailed design phase. Certain very large, high-pressure valves with delivery dates exceeding one year may be ordered during the FEED stage of the project.

The procurement cycle involves the following activities:

- Locate suitable approved vendors.

- Make requests for quotation (RFQs).

- Receive and distribute bids for the technical bid evaluation (TBE), which is carried out by the piping material engineer responsible for the package.

- Prepare a commercial evaluation.

- Hold bid clarification meetings.

- Place purchase orders.

- Hold a kickoff meeting with the vendor.

- Commence commercial negotiations for any deviations or delivery delays.

- Establish control all vendor documents and prepare a distribution matrix.

Purchase orders are issued, followed by an inspection plan to monitor the quality of the component during the various manufacturing phases up to transportation to the site. The inspection activities are specified in an inspection test plan, a document that outlines the following:

- The roles and responsibilities for the owner, EPC contractor, technical authority, and third-party inspection contractor, if applicable.

- Description of the inspection activity.

- Applicable technical documents.

- Acceptance criteria.

- Certification requirements.

During the manufacturing process, the progress of the component must be monitored by the Expediting Department, which is under the control of the procurement group. Its responsibility is to ensure that the contractually agreed-on required-on-site (ROS) date is not missed. If any slippage is expected, then site and the construction groups must be made aware, immediately, of any delays in delivery, to allow rescheduling construction activities, if necessary. The Procurement Department also must be aware of any delays, because they may have a cost impact on the purchase order and penalty clauses may have to be applied against the manufacturer.

5.2.5 Construction Phase

The construction phase of a project usual starts midway through the detailed design phase, and it starts with the civil and structural engineering disciplines. The major activities during this phase are

- Site clearance (civil engineers).

- Roads (civil engineers).

- Trenching for underground piping (civil engineers).

- Foundations for major equipment (civil engineers).

- Pipe racks and pipe tracks (structural engineers).

The piping content for the construction phase generally starts approximately 75% through the detail design phase, after the P&IDs are issued for construction. The 3D modeling is in progress and 30%, 60%, and 90% of the model reviews should be completed and the extraction of the isometric IFC commenced.

The fabrication of piping spool usually takes place at the job site or a location very close to it, to reduce the transportation costs and simplify the logistics. IFC isometrics are issued by the EPC contractor responsible for the detailed design and sent to the piping fabricator, who is often a subcontractor.

Advances in information technology mean that these production isometrics can be sent electronically to the fabrication shop, where they can be converted into spool isometrics. The fabrication site will have been receiving piping material for several months, and it is essential that there is harmony between the delivery of material, delivery of isometrics, and the construction sequence, because

- It is pointless to issue isometrics if the material has not arrived.

- It is pointless if the material has arrived and the fabrication shop is waiting for the isometric to be issued.

Assume all the piping material and the isometrics have arrived and the spool is fabricated but it is not in line with the construction sequence. For a certain period this can be acceptable, because spools can be stored for a period of time, but there is a limit to the space available. The ideal situation is that, once a spool has been fabricated and passes all the nondestructive examination tests, it should be transported to site immediately and installed as quickly as possible, in line with the construction sequence.

As the 3D model approaches 100%, the isometric extraction continues until the project is complete.

Hydrotesting the erected pipe usually follows the construction sequence. As the piping system becomes complete, it can be subjected to a hydrotest in accordance with the design code for the project. Normally, this is 1.5 times the pressure class rating at ambient for a hydrotest and 1.1 for a pneumatic test. The air test is lower, because of the high danger level with the stored energy of the compressed air contained within the piping system.

Piping systems in a specific unit or area are tested until all the systems in the unit or area are mechanically complete and the project can move to the commissioning phase. Mechanical completion occurs when

- All design and engineering has been completed.

- All equipment has been completed in accordance with the IFC drawings, specifications, applicable codes, and regulations.

- All instruments have been installed.

- All factory acceptance tests and all other testing and inspection activities have been completed.

- The EPC contractor has obtained all approvals for which it is responsible.

- All hazardous operations checks have been completed.

- Operating procedures and maintenance procedures are on site.

- Third-party, regulatory, and company approval have been obtained and confirmation documentation provided to the company.

By the middle of the construction phase, the detailed design group will be gradually reduced in size, and it is not uncommon for certain key personnel to be transferred to the site as field piping engineers, so that they can use their experience from the detailed design stage to answer site queries.

For the piping group, the major deliverables during this phase are as follows. The piping material engineering group prepares

- Review and approval of vendor documents.

- Review and approval of deviations from manufacturers.

- Resolution of site material queries.

The piping design group prepares a piping CAD 3D model and the piping isometrics (IFC). The piping stress group prepares a Piping flexibility analysis for critical lines (IFC) and the stress isometrics (IFC) and assembles stress packs (IFC). The piping material control group prepares top-up material takeoffs for requisitions (IFC) and the control reports of piping materials delivered to the site (IFC).

5.2.6 Precommissioning and Commissioning Phase

The purpose of the precommissioning phase, after the project is considered to be mechanically complete, is to prepare the facility for commissioning. It can include the following piping activities:

- Calibration checks of instruments and loop checking.
- Testing of PSVs.
- System flushing and cleaning.
- System drying (where applicable).

When the commissioning manager is satisfied that the precommissioning activity has been successfully completed, commissioning can commence. Commissioning involves testing the integrated system within a unit or area, which includes piping and all the related items of equipment: usually the test media is the final product that the systems will carry during plant operation. Commissioning is an ongoing activity that moves around a unit or area until it is complete.

Commissioning can start at the tail end of the detailed design phase, when lead piping engineers and key personnel are still working on the project and are available to answer any technical questions.

There are cases when the detailed engineering phase is at a closeout stage, key personnel may be off the job, and only lead piping engineers are left.

The deliverables from the piping group during the commissioning phase are limited to specific isometrics for site modifications, deviations for piping materials still on order, and site queries.

5.2.7 Startup and Handover to the Owner

When all the units and areas have been successfully commissioned, before startup and handover to the operator/end user, the plant has to successfully complete a performance test for a fixed period of time.

At this stage of the proceeding, the detailed design phase for the piping group will have been closed out, and the only personnel left in the design office will be project engineers.

CHAPTER 6

Fabrication, Assembly, and Erection

A piping system is made up of numerous assemblies and subassemblies that have to be fabricated, inspected, installed, and tested to form a completed system. Each step is of equal importance, and in this chapter, I deal with the fabrication and installation of piping systems.

The *fabrication* of pipe is the bending, cutting, forming, and welding of individual pipe lengths and piping components to each other and, if specified, the subsequent heat treatment of subassemblies.

The *erection* or *installation* of pipe refers to placing the fabricated piping assemblies, valves, and other special piping items in their final location to interconnect pumps, compressors, heat exchangers, turbines, boilers, and other items of process equipment. The fabrication of a piping assembly can be carried out in one of the following locations:

- Commercial fabrication shop distant from the job site, shop fabrication.
- A fabrication shop on the job site, shop fabrication.
- On site close to or at the final site for installation, site fabrication.

The decision on the location of the fabrication shop is based on cost, logistics, and availability of qualified personnel. A commercial fabrication shop has specialized equipment, such as automatic welding machines or large-diameter pipe bending machines, that might not be available at a fabrication facility on site. Shop fabrication is carried out under controlled conditions and in a predictable environment; and it is the preferred location for fabrication. Site fabrication means a limit to the availability of special machines; and it is carried out at the mercy of local weather conditions, however, sometimes site fabrication is unavoidable.

Fabrication of piping subassemblies generally is carried out in a dedicated shop for pipe NPS 3 (DN 75) and larger. Piping NPS 2 (DN 50) and smaller usually are field fabricated in the final location, unless it requires special welding, inspection, or an internal cleaning process. In cases where there are long straight runs of pipe, like those on pipe racks or for pipelines, these pipe sections are welded together in the field very close to their final location.

6.1 Codes and Standards Considerations

Many codes and standards are used for to the fabrication, erection, and inspection of process piping systems, some of them are mandatory. The code I refer to most in the book is ASME B31.3; and the subject of fabrication, assembly and erection is covered in Chapter 5 of this code.

6.2 Fabrication Materials for Piping Systems

Carbon steel is the most commonly used material of construction used for process piping systems, followed by stainless steel and various alloys. Many nonmetallic materials also are used. Material are selected according to their corrosion resistance to the fluid and ability to handle the design temperature and design pressure. Piping systems are fabricated from a great variety of metals and nonmetals, material selection being a function of the environment and service conditions.

The material used for fabrication must conform to a relevant ASTM, API, or other recognized standard that guarantees the predictability of

- Chemical composition.
- Mechanical properties.
- Physical properties.
- Heat treatment.

When the materials of construction are known, the fabricator can select the correct welding procedure to ensure the best possible weld and a leak-free joint under the pressure and temperature conditions the piping system will be subjected to in service.

ASME B31.3 has a list of materials that are acceptable for the construction of piping systems designed to that code. This does not exclude the use of other materials: but before an unlisted material is used, it must be qualified for use.

6.3 Fabrication Drawings

The EPC contractor is responsible for creating the 3D piping model using the IFC P&IDs as the basis. From the model, the production piping isometrics are extracted. These issued for construction isometrics are considered to be the basis of the fabrication of a piping system. An isometric is a sketch that, although not to scale, carries all the dimensional and material information necessary to fabricate and erect a piping system. An isometric contains piping material that will be fabricated in the shop and field material and the location of any field welds that will be used to erect the piping system in situ.

These IFC isometrics then are sent to the piping fabricator, who divides this piping assembly into subassemblies, which contain only materials that have to be fabricated. These subassemblies are called *spool isometrics*. The size of these spools is based on a number of factors:

- Overall dimensions of the spool.
- Field welds.
- Weight.
- Heat treatment.
- Availability of space at the fabricator.

The spool isometrics carry all the necessary dimensions, angles, and make allowances for all component and fabrication tolerances specified in the relevant codes and standards. Piping spool tolerances normally conform to PFI-ES-3, "Fabricating Tolerances."

6.4 Fabrication Activities

Numerous types of activities take place in the fabrication shop, but these are the most significant:

- Cutting.
- Beveling.
- Forming.
- Bending.
- Welding.
- Brazing and soldering.
- Galvanizing.

The two most basic fabrication activities are cutting and beveling.

6.4.1 Cutting

The most fundamental fabrication activity, it applies almost exclusively to pipe lengths, because pipe fittings are supplied in standard dimensions and the 3D modeler uses these dimensions to create a piping system. The isometric uses these standard component dimensions to create isometrics, from which spools are created. The cutting method generally is mechanical or thermal. The mechanical methods, or cold cuts, involve the use of saws, abrasive discs, and pipe cutting machines. The thermal methods, also known as *hot cuts*, involve flame cutting using gas or electric arc cutting.

6.4.2 Beveling

After a pipe length has been cut, the end is at 90° to the axis of the pipe and further preparation is required before welding can take place. Beveling is the process to profile the correct shape at the end of a pipe to allow it to be mated to the other pipe or piping component and creates a groove that can be welded. This can be a single bevel or, on thicker walled pipes, a compound bevel at two or more angles. The beveling can be completed by either mechanical or thermal methods.

6.4.3 Forming

Forming is piping fabrication that includes bending, extruding, swaging, lapping, and expanding to create a component of a connection. The standardization of welded pipe fitting has meant that this fabrication process for components is not commonly required; however, it is an option that ASME B31.3 allows.

6.4.4 Bending

The cold or hot bending of straight pipe is another option and a more commonly used method of fabrication than forming. Small-bore piping usually less than 2" for utility services can be bent if an approved procedure is applied. Also, large pipelines lines that required to be pigged for cleaning or batching purposes, require long radius bends of three and sometimes five times the outside diameter to allow a smooth passage for the pig. These can be bent from pipe. Bending is more common on pipeline projects designed to ASME B31.4, Pipeline Transportation Systems for Liquid, and ASME B31.8, Gas Transportation and Distribution Piping Systems.

Three very important dimensional limitations must be addressed when considering bending pipe: thinning, buckling, and ovality. That is why a comprehensive bending procedure is required and a suitable bending method chosen.

Thinning is important because, during the bending operation, the outer edge is stretched and the inner edge compressed. The stretching causes a thinning of the wall thickness of the outer section, which has to be monitored and should not exceed the tolerance allowable for the pipe.

Buckling is important because the bending operation has an opposite effect on the inner wall thickness and has the tendency to compress; however, this does not always result in a thickening of the wall section and there is a tendency, at a certain stage of compression, for the inner edge to buckle.

Ovality is important because, during the bending operation, the cross section of the bend can assume a oval shape. The degree of ovality is determined by the difference between the major and minor axes divided by the nominal diameter of the pipe. ASME B31.3 states that the difference between maximum and minimum diameters at any cross section shall not exceed 8% of nominal outside diameter for internal pressure and 3% for external pressure. Removal of metal shall not be used to achieve these requirements.

The cold bending of ferritic materials should be done at temperatures below their transformation range. Hot bending should be done at a temperature above their transformation range. After the cold or hot bending process, heat treatment is performed in accordance with ASME B31.3, paragraph 331.1.1.

Other significant subjects regarding the fabrication and installation of process piping systems are covered in ASME B31.3, Chapter 5, "Fabrication, Assembly, and Erection," which covers the following topics.

Welding

- Welding responsibility.
- Welding qualifications.
- Welding materials.
- Preparation for welding.
- Welding requirements.
- Weld repair.

Preheating

- General.
- Specific requirements.

Heat Treatment

- Heat treatment requirements.
- Governing thickness.
- Heating and cooling.
- Temperature verification.
- Specific requirements.
- Alternative heat treatment.
- Exceptions to basic requirements.

- Dissimilar materials.
- Delayed heat treatment.
- Partial heat treatment.
- Local heat treatment.

Bending and Forming

- Bend flattening.
- Bending temperature.
- Corrugated and other bends.
- Forming.
- Required heat treatment.
- Hot bending and forming.
- Cold bending and forming.

Brazing and Soldering

- Brazing qualification.
- Brazing and soldering materials.
- Filler metal.
- Flux.
- Preparation.
- Surface preparation.
- Joint clearance.
- Requirements.
- Soldering procedure.
- Heating.
- Flux removal.

Assembly and Erection

- Alignment.
- Flanged joints.

- Preparation for assembly.
- Bolting torque.
- Bolt length.
- Gaskets.
- Threaded joints.
- Thread compound or lubricant.
- Joints for seal welding.
- Straight threaded joints.
- Tubing joints.
- Flared tubing joints.
- Flareless and compression tubing joints.
- Caulked joints.
- Expanded joints and special joints.
- General.
- Packed joints.
- Cleaning of piping.

6.5 Welding

Welding plays a very important part in the fabrication of process piping systems, and it is essential that the correct procedures and suitable, qualified welders are used. ASME B31.3, Process Piping, references ASME Section IX as the minimum requirements for qualifying welding procedures (WPS, or welding procedure specifications) and welding personnel. A well-defined WPS that references the base metal, filler material, shielding fluxes/gases, positions, and heat treatment can produce a welded joint with the required characteristics and leak free. The parameters are recorded in a procedure qualification record (PQR).

Fabrication must be executed by personnel who are qualified to work to the relevant WPS. According to ASME Section IX rules, a qualified welder who has not welded in a specific WPS within a specified period of time must be requalified.

6.5.1 Welding Processes

The welding processes most commonly used for the fabrication of process piping systems are

- Shielded metal arc welding (SMAW).
- Submerged arc welding (SAW).
- Gas tungsten arc welding (GTAW).
- Gas metal arc welding (GMAW).

However, any welding process that can be qualified under the requirements of ASME Section IX is acceptable.

Shielded Metal Arc Welding.

The shielded metal arc manual welding process uses the heat of the arc to melt the base metal and the tip of a consumable covered electrode. The electrode is flux coated with a metal filler. The electrode forms part of the electric circuit when the arc is struck. This circuit begins with the electric power source and includes the welding cables, an electrode holder, a work piece connection, the work pieces to be welded together, and an arc welding electrode. One of the two cables from the power source is attached to the work. The other is attached to the electrode holder.

Welding commences when an electric arc is struck by making contact between the tip of the electrode and the work piece. The intense heat of the arc melts the tip of the electrode onto the surface of the work piece adjacent to the arc.

The molten metal rapidly forms at the tip of the electrode, and it is transferred into the molten weld pool. The filler metal is deposited as the electrode is progressively consumed and the flux protects the welding process from oxidization. The arc moves across the work piece at the appropriate arc distance and travel speed, melting and fusing a portion of the base metal and continuously adding filler metal.

Submerged Arc Welding

SAW is an automatic or semiautomatic welding process best suited for circumferential butt-welds on pipe in the horizontal position. An arc is created between the work piece and a bare solid wire, which is consumed during the welding operation. The electrode is in a coil, and the shielding is carried out by a blanket of granular, fusible material, called the *flux*, which covers both the arc and molten metal by forming a slag blanket to prevent oxidation. It acts as a thermal insulator, permitting deeper heat penetration. SAW deposits weld at the greatest rate, which results in very high welding productivity, 4–10 times as much as the shielded metal arc welding, and it is the preferred procedure.

Gas Shielded Arc Welding

The term *gas-shielded arc welding* refers to welding processes where the arc and molten metal are shielded from oxidation by some type of inert gas rather than by a flux.

Gas Tungsten Arc Welding

Gas tungsten arc welding, also known as *tungsten inert gas* (TIG) *welding*, is an arc welding process that uses a nonconsumable tungsten electrode to produce the weld. The filler metal is added from an external source, usually as bare metal filler rod. The weld pool area is protected from the atmosphere and possible contamination by a shielding inert gas, such as argon. A filler metal normally is used, although some welds, known as *autogenous welds*, do not require it.

GTAW is best suited to weld thin sections of stainless steel and light metals, such as aluminum, magnesium, and copper alloys. The process allows the operator greater control over the welding process than other methods, which results in stronger, high-integrity welds. The disadvantages are that GTAW is more complex and slower than many other welding techniques.

Gas Metal Arc Welding

GMAW, also known as *metal inert gas* (MIG) *welding*, is a type of gas-shielded welding generally used in the manual mode but can be automated. The filler wire also is the electrode, and it is supplied in coils of solid bare wire. The coil is fed automatically into the joint, melted in the arc, and deposited in the weld groove. Alloying elements arc in the wire, and the shielding inert gas may be argon, helium, nitrogen, carbon dioxide, or a combination of these gases, depending on the application.

Weld Repair

Any weld considered to be defective and requires repair must be ground back to the base metal. The weld under repair must use the correct WPS, bearing in mind that the contour surface may be different in profile and dimensions from the original. The preheating and heat treatment should be the same as specified for the original welding. The two most common types of welds used in a fabrication shop are but-weld and socket weld. These two types of weld are not restricted to the fabrication shop. Both welds are used in the field, on site: but because the shop has access to more equipment and a controlled environment, if logistically possible, the shop is the first choice to complete a weld. In many cases, this is not possible.

The butt-weld requires special end preparations, and it can be used on pipe of all commercial sizes. The socket weld is meant only for pipe up to NPS 4 (DIN 100), although in reality, socket weld fittings are rarely specified above NPS 2 (DIN 50).

6.5.2 Preheating

Preheating before the welding process helps slow the cooling rate of the weld joint, which results in a high level of ductility in the final weld and the heat-affected zone (HAZ). It allows dissolved hydrogen to diffuse more readily, which helps reduce shrinkage, distortion, and possible cracking caused by residual stresses.

Different base metals in WPS have different required and recommended preheat temperatures, and these are specified in ASME B31.3, Table 330.1.1, Preheat Temperatures.

6.5.3 Heat Treatment

Heat treatment is used to avoid or alleviate the detrimental effects of high temperatures, which are unavoidable during the welding process. The heat treatment required and the recommended temperatures and holding times vary among metals. They are specified in ASME B31.3, Table 331.1.1, Requirements for Heat Treatment, except were specified differently in the same document. Materials are grouped under P numbers, which originate from BPV Code, Section IX, QW/QB-422. When using this table, generally, the thicker of the two components is considered to be the governing thickness when specifying heat treatment.

The method of heating the work piece should be applied uniformly under a controlled condition, which includes an enclosed furnace, local flame, electric induction, electric resistance, or any other approved method. The cooling-down method also must be in a controlled environment, which allows uniform temperature reduction.

At all stages of the heat treatment process, the temperature must be accurately monitored and recorded by a thermocouple or other suitable, approved method, to guarantee that the temperatures specified in the WPS are adhered to during the welding process. When necessary these temperatures must be recorded.

Where hardness tests of production welds are stipulated to verify satisfactory heat treatment, the hardness limit applies to the weld and to the heat affected zone tested as close as practicable to the edge of the weld.

The heat treatment of welded joints between dissimilar ferritic metals or between ferritic metals using dissimilar ferritic filler metals should use the higher temperature specified in ASME B31.3, Process Piping Table 331.1.

If a weld is allowed to cool prior to the postweld heat treatment, the rate of uniform cooling must be controlled to prevent a detrimental effect in the piping.

Partial heat treatment of a piping assembly that cannot be fitted completely into a furnace is acceptable if there is at least a 300 mm (1 ft) overlap between successive heats and parts of the assembly outside the furnace are protected from harmful temperature gradients.

The local heat treatment of a weld is acceptable if the heat treatment is applied circumferentially on the run pipe and the branch. This section should be heated until the specified temperature exists over the entire pipe section and at least 25 mm (1 in.) beyond the section.

6.6 Brazing and Soldering

The qualification for brazing procedures, brazing equipment and brazing operators is in accordance with the requirements of the BPV code, Section IX, for category D fluid service at design temperatures not over 93°C (200°F).

The brazing alloy or solder should have a melting temperature within the specified or desired temperature range and, in conjunction with a suitable flux or controlled atmosphere, wet and adhere to the surfaces to be joined.

A flux that is fluid and chemically active at the brazing or soldering temperature is used when necessary to eliminate oxidation of the filler metal and the surfaces to be joined and to promote free flow of brazing alloy or solder.

Prior to brazing or soldering, all surfaces in contact with the operation are made clean and free from grease, oxides, paint, scale, and dirt of any kind. A suitable chemical or mechanical cleaning method is used to provide a dirt-free surface that is not water resistant. The fit-up clearance between surfaces of the two components to be joined by brazing or soldering should not be so large as to prevent complete capillary distribution of the filler metal. Soldering should follow the procedure in the *Copper Tube Handbook* of the Copper Development Association.

To minimize oxidation, the joint is brought to brazing or soldering temperature in as short a time as possible without localized underheating or overheating. After the soldering operation, all residual flux is removed.

6.7 Protection of Carbon Steel in Corrosive Services

The following are methods of giving basic carbon steel longer in-service life when it is subjected to a corrosive service:

- Adding an additional corrosion allowance, usually 1.5 mm, 3 mm, or 6 mm to the calculated wall thickness of the pipe or component.
- Internally galvanizing the pipe and piping systems.

- Internally cladding the carbon steel pipe with a corrosion-resistant alloy.

6.7.1 Corrosion Allowance

Carbon steel can be used for a piping system even if there is a possibility of the service fluid causing some corrosion to the carbon steel pipe. A weight loss study is carried out to determine the volume of pipe material that will be lost to corrosion over the intended in-service life of the piping system. This weight loss is converted into an additional internal wall thickness added to the wall of the pipe, which has been calculated for pressure containing purposes, and called a *corrosion allowance* (CA).

6.7.2 Internal Galvanizing of Pipe and Piping Systems

Another method of arresting corrosion of piping systems in certain utility services operating at ambient temperatures is to galvanize them with an internal zinc based coat.

Pipe galvanizing requires four steps: cleaning, fluxing, galvanizing, and degreasing. A galvanizing procedure must be followed to ensure that the galvanized coat is of a high quality and unlikely to be removed.

Carbon steel can be galvanized effectively only when its surface has been chemically cleaned using a specified solvent; and the surfaces are free of contaminants, such as oil and grease, that will have an adverse effect on the galvanizing process.

Oil and grease can be removed with caustic solutions, which usually form part of the cleaning process; however, pickling in acids, like hydrochloric or sulfuric, is the most widely used method of removing millscale and oxides from steel pipe. Pickling cleans the pipe surfaces and removes millscale and rust. The surfaces will reoxidize if left in unprotected, ammonium chloride, a flux widely used by pipe galvanizers. Zinc ammonium chloride serves two very important purposes: it prevents additional oxide from forming on the surfaces of pickled pipes, and it helps molten zinc react with these surfaces during the galvanizing process.

Pipes that have been fluxed should be dry before they enter the zinc bath; otherwise, some areas may fail to react successfully with the zinc and remain ungalvanized. These areas are known as *black spots*. For this reason, the flux must be carefully controlled and its temperature and concentration levels must be monitored and adjusted to achieve the best results.

The galvanizing process takes place in either a galvanizing kettle or a galvanizing bath. Galvanizing baths can be heated by gas, oil, or electricity. The energy requirements should be sufficient that sustained temperatures can be reached initially to melt the zinc feedstock and keep it molten for the duration of the galvanizing process. Oil and gas heating probably are the most common; and heat is applied via burners along the side of the bath, with care taken that the heat delivered is uniform and has no hot or cold spots. Zinc ingots of approximately 25 kg are added to the heated bath.

The galvanizing reaction takes place within the temperature range of 440°C to 470°C (825°F to 878°F), with the nominal bath temperature of 450°C (840°F). Zinc solidifies at around 419°C (786°F), the minimum operating bath temperature that must be maintained for the process to be successful.

Pipes that have been chemically cleaned and fluxed are placed in the bath and fully immersed to ensure that the internal surfaces are covered by the molten zinc. After being in the bath for the predetermined time, the pipes are individually retrieved from the molten zinc. The time that the pipes are in the bath should be sufficient to raise their temperature to above 440°C (824°F), at which point galvanizing occurs at a reasonable rate. The desired coating weight also is a factor. Immersion time, typically 2–4 minutes, is governed automatically by controlling the speed of the pipe galvanizing machine.

Chemically cleaned and fluxed pipes enter the zinc bath along the side of the kettle. Two designs of pipe galvanizing machinery are widely used. The main difference between them is the way pipes are submerged in the zinc bath. The pipes should be pushed into the zinc bath at a slight angle to allow air and steam to escape freely. In addition to being an important safety feature, putting pipes into the kettle at an angle often improves the quality of the galvanized coating on the inside of pipes. Typically, several pipes are allowed to roll from a loading table into the bath at one time, often with manual assistance. Once in the bath, a mechanism lowers the pipes at a predetermined rate. This mechanism can be a pusher rod or a notched wheel. After residing in the bath for the prescribed time, the pipes are individually retrieved from the molten zinc.

The required thickness is achieved by passing each pipe through a perforated ring blowing superheated air onto the outside of each pipe upon removal from the galvanizing bath. Pressurized steam is used to remove excess zinc from the inside of each pipe.

The ASTM standards that are referred to with regard to galvanizing of pipe and piping systems are in Table 6–1.

6.7.3 Pipe Cladding

Another method of protecting the carbon steel base material from a corrosive fluid is to internally clad the piping system with a corrosion-resistant alloy metal such as stainless steel, alloy 625, or alloy 825 or to line it with a nonmetallic material such as concrete, PTFE, glass, or epoxy. The choice of cladding or lining a piping system is based on

- Economics.
- A proven cladding procedure.
- A qualified cladding fabricator.
- Additional inspection.

The decision to use a cladded piping system should be made very carefully, because it introduces an additional activity that requires careful monitoring to ensure the integrity of the cladding and subsequent fabrication.

6.8 Assembly and Erection

6.8.1 Alignment

A degree of misalignment is acceptable, in ASME B31.3, for the mating of assemblies to complete a piping system; however, they must not introduce strain that will be detrimental to the components or equipment. Cold spring of process piping systems is not prohibited in ASME B31.3; however, most operators and EPC contractors avoid this option. For a flanged joint bolt-up, flange faces should be aligned in the design plane within 1 mm in 200 mm (1.16 in./ft) measured

6.8 Assembly and Erection 187

Table 6–1 ASTM Standards Relating to Hot-Dip Galvanizing and Hot-Dip Galvanized Materials

A 36	Specification for structural steel
A 123/A 123 M	Specification for zinc (hot-dip galvanized) coatings on iron and steel products
A 143	Practice for safeguarding against embrittlement of hot-dip galvanized structural steel products and procedure for detecting embrittlement
A 153/A 153 M	Specification for zinc coating (hot-dip) on iron and steel hardware
A 384	Practice for safeguarding against warpage and distortion during hot-dip galvanizing of steel assemblies
A 385	Practice for providing high-quality zinc coatings (hot-dip)
A 500	Specification for cold-formed welded and seamless carbon steel structural tubing in rounds and shapes
A 501	Specification for hot-formed welded and seamless carbon steel structural tubing
A 563	Standard specification for carbon and alloy steel nuts
A 572	Specification for high-strength low-alloy columbium-vanadium steels of structural quality
A 767/A 767 M	Specification for zinc-coated (galvanized) steel bars for concrete reinforcement
A 780	Practice for repair of damaged and uncoated areas of hot-dip galvanized coatings
A992	Specifications for steel structural shapes for use in building framing
B 6	Specification for zinc
D 6386	Practice for preparation of zinc (hot-dip galvanized) coated iron and steel products and hardware surfaces for painting
E 376	Practice for measuring coating thickness by magnetic-field or eddy-current (electromagnetic) test methods

across any diameter; flange bolt holes should be aligned within 3 mm (1.8 in.) maximum offset.

6.8.2 Flanged Joints

Flange faces to be bolted up must be parallel, free from damage, and have a surface type and finish that is acceptable for the gasket in the specifications.

The most commonly specified flange types are

- Weld-neck flange.
- Socket-weld flange.
- Screwed flange.
- Slip-on flange.
- Lap-joint flange with a stub end.

The most commonly specified flange faces are raised face, ring-type joint, and flat faced. The most commonly used standard that covers dimensional information on flange types and flange faces used in process piping systems design to ASME B31.3 code is ASTM B16.5.

An approved bolting procedure must be employed to ensure that the gasket is uniformly compressed to the recommended design loadings to achieve a leak-free joint. Special care must be taken with bolt-ups of flanges of different materials or differing mechanical properties.

Bolt lengths must be specified so that they extend completely through the nut and full thread engagement is achieved. ASME B31.3 allows a lack of complete engagement of one thread.

Only one gasket should be used between two flanged joint assemblies.

6.8.3 Threaded Joints

Threaded piping components are available up to NPS 4 (DIN 100), although in reality, threaded fittings are rarely specified above NPS 2 (DIN 50) and restricted to low-pressure, nonhazardous utility services at ambient temperatures.

For threaded joints, the compound or lubricant used on the mating surfaces must be compatible with the piping component and suitable for the fluid and the design service conditions.

A threaded joint that will be welded should be made up without a thread compound to avoid the possibility of creating a defective weld.

CHAPTER 7

Inspection and Testing

7.1 Piping Codes

Numerous industry codes reference the inspection and testing of piping systems and standards relating to piping systems, and these activities are required by the authorities and insurance companies, but the two codes most commonly referenced internationally for process piping systems are the ASME B31 Codes for Pressure Piping and the ASME Boiler Pressure Vessel Code.

The inspection, examination, and testing of piping systems is covered in Chapter VI of B31.3, and Chapter VI covers the following subjects:

Inspection

- Responsibility for inspection.
- Rights of the owner's inspector.
- Qualifications of the owner's inspector.

Examination

- Responsibility for examination.
- Examination requirements.
- Acceptance criteria.
- Defective components and workmanship.

- Progressive sampling for examination.
- Extent of required examination.
- Examination normally required.
- Examination—category D fluid service.
- Examination.
- Supplementary examination.
- Spot radiography.
- Hardness tests.
- Examinations to resolve uncertainty.

Examination Personnel

- Personnel qualification and certification.
- Specific requirements.

Examination Procedures

Types of Examination

- Methods.
- Special methods.
- Definitions of examination.
- Visual examination.
- Magnetic particle examination.
- Liquid penetrant examination.
- Radiographic examination.
- Method of radiography.
- Extent of radiography.
- Ultrasonic examination.

- Method of ultrasonic.
- Acceptance criteria.
- In-process examination.

Testing

- Required leak test.
- General requirements for leak tests.
- Limitations on pressure.
- Special provisions for testing.
- Externally pressured piping.
- Jacketed piping.
- Repairs or additions after leak testing.
- Test records.
- Preparation for leak test.
- Joints exposed.
- Temporary supports.
- Piping with expansion joints.
- Limits of tested piping.
- Hydrostatic leak test (test fluid, test pressure, and hydrostatic test of piping with vessels as a system).
- Pneumatic leak test (precautions, pressure relief device, test fluid, test pressure, and procedure).
- Hydrostatic-pneumatic leak test (initial service leak test, test fluid, procedure, and examination for leaks).
- Sensitive leak test.
- Alternative leak test.

- Examination of welds.
- Test method.

Records

- Responsibility.
- Retention of records.

Inspection (the ASME B31.3 code distinguishes between examination and inspection)

- Examination applies to quality control functions performed by the manufacturer of piping components for the fabricator or erector.
- Inspection applies to functions performed for the owner by the owner's inspector or the inspector's delegates.

The owner of the plant has the responsibility to verify that all the required examinations and testing have been completed and to inspect that all the process piping systems conform to all the applicable examination requirements of the relevant code. These activities are carried out by the owner's inspector or the inspector's delegates.

The owner's inspector has the right of access to any location where work concerned with the piping fabrication or installation is being performed. These locations include the manufacture, fabrication, heat treatment, assembly, erection, examination, and testing of the piping.

The owner's inspector acts independently and cannot be an employee of the piping manufacturer, fabricator, or erector unless the owner is also the manufacturer, fabricator, or erector. The inspector must have as a minimum 10 years experience in the design, fabrication, or inspection of industrial pressure piping.

If any of this inspection activity is delegated to a third party, the owner's inspector is responsible for determining that the individual to whom an inspection function is delegated is suitably qualified to perform that function.

Inspection by the owner's representative does not relieve a manufacturer, fabricator, or the erector of the responsibility for providing materials, components, and workmanship in accordance with the requirements of the code or performing all required examinations and preparing suitable records of examinations and tests.

Prior to plant startup, each piping installation, including components and workmanship, must be examined to a level in accordance with the code requirements. The acceptance criteria for inspection is stated in the engineering design and it must meet the applicable requirements stated in appropriate code.

Defective components or poor workmanship discovered during examination that exceed the acceptance criteria of this code must be repaired or replaced. The new work must be reexamined using the same methods, to the same extent, and by the same acceptance criteria as required for the original work. If a spot or random examination reveals a defect in a component or workmanship, it is necessary to increase the level of examination to the same kind of component or work activity carried out by the same operator.

Piping systems for normal fluid service should be examined to the extent specified in the relevant design code or to greater engineering design if it is more stringent. The types of examination include visual, DPI or MPI, and radiography or ultrasonics.

The type of examination and the extent, along with the results, must be documented and all material certification and data sheets for piping components retained to give evidence of chemical composition, mechanical properties, relevant heat treatment, and dimensional data.

Piping systems for category D fluid service, as designated in the engineering design, should be visually examined in accordance with the relevant code to satisfy the examiner that piping components, materials, and workmanship conform to the necessary requirements.

Piping systems for severe cyclic conditions should be examined to the extent specified in the relevant code or to the engineering design, if more stringent. Acceptance criteria are stated in the relevant code.

Spot radiography must be carried out on the following types of fabrication:

- *Longitudinal welds.* Spot radiography for longitudinal groove welds required to have a weld joint factor of 0.90 requires examination by radiography in accordance with the code.
- *Circumferential butt-welds and other welds.* It is recommended that the extent of examination be no less than one shot on 1 in each 20 welds for each welder.

All examination personnel should be suitably trained and have the relevant experience necessary to complete this activity to the satisfaction of the owner. Resumes and qualifications must be maintained and available to the owner's inspector. Examination procedures must be performed in accordance with a written procedure that conforms to one of the methods specified in the relevant code and the engineering design.

7.2 Types of Examination

Any examination method specified by the code, the owners inspector, or the engineering design must be performed in accordance with one of the methods specified in the following list. Special examination methods that are not specified in a code but are required by the owner's inspector or in the engineering design must be clearly specified in sufficient detail and the acceptance criteria defined to permit qualification.

The following terms apply to the extent of any type of examination:

- *100% examination.* This means the complete examination of all of a specified kind of item in a designated lot of piping components or fabrication work.
- *Random examination.* This is a complete examination of a percentage of a specified kind of item in a designated lot of piping.
- *Spot examination.* This is a specified partial examination of each of a specified kind of item in a designated lot of piping,

such as part of the length of all shop-fabricated welds in a lot of jacketed piping.

- *Random spot examination.* This means a specified partial examination of a percentage of a specified kind of item in a designated lot of piping.

7.2.1 Visual Examination

The visual examination of all or part of a piping system must be performed in accordance with the code, such as referenced in BPV Code, Section V, Article 9. Visual examination is carried out by the naked eye and refers to defects in components and workmanship without the aid of measuring or recording equipment. Visual inspection the first option and most commonly used method of nondestructive testing.

Visual examination is used for the following applications:

- Correct weld preparation.
- Surface defects, such as undercuts and cracks.
- Weld profile (cap, root).
- Measurement of component geometry.
- Measurement of surface roughness.
- Detection of corrosion.
- Detection of defects and cracks.
- Detection of excess grinding defects.

7.2.2 Liquid Penetrant Examination

Liquid penetrant examination of welds and of components other than castings should be performed in accordance with BPV Code, Section V, Article 6.

After visual inspection, liquid penetrant testing is probably the oldest and the second most commonly used method of NDT. It can be used on any nonporous material, however, its use is restricted to the detection of surface defects.

A colored or fluorescent dye is applied to a cleaned surface of the work piece and allowed to settle for between 10 and 20 minutes. This dye is drawn into the discontinuity by capillary pressure and penetrates the surface. This capillary pressure is determined by the width of discontinuity, surface tension, and contact angle of the dye on the surface.

Excess dye is then removed from the surface and a developer applied, which draws out the dye and gives good visual contrast to the defect. The work piece to be inspected is left for approximately 10 minutes, then the surfaces are visually inspected, using a white light for red dyes and ultraviolet light for fluorescent dyes.

7.2.3 Magnetic Particle Examination

Magnetic particle examination (MPE) of welds and components other than castings should be performed in accordance with the code, such as referenced in BPV Code, Section V, Article 7.

MPE is a very effective method for the detection of surface and close-to-the-surface discontinuities in any ferromagnetic material. It relies on the principle that a magnetic field is uniform through a component unless disturbed by the presence of a flaw. A flaw generates a local stronger field, known as a *leakage field*. It attracts finely ground magnetic particles, either as an ink or dry powder, which are applied to the component, providing a visual indication, making the defect visible.

A number of factors can influence the production and quality of indications: defect size, depth of defect below the surface, magnetizing, current level of magnetization relative to the orientation of the defect, direction of magnetization, magnetic properties of test piece size, and magnetic properties of detecting media.

The advantages to this method are

- Simple examination.
- Ability to detect surface and near-surface flaws.

- Ability to detect flaws filled with contaminants, such as oxide or nonmetallic inclusions.

The disadvantages are

- Application limited to ferromagnetic materials.
- Inability to detect deep internal flaws.
- Possibility of damage due to high currents applied to the component.
- Components usually have to be demagnetized.

7.2.4 Radiographic Examination

Radiography of welds and components other than castings should be performed in accordance with the code as referenced in the code, such as BPV Code, Section V, Article 2.

Radiography is an accurate examination method that uses X-ray equipment in the laboratory and radioisotopes on site. It is the most costly of all of the examinations mentioned in this section and produces a permanent record of the work piece being examined. Radiography is used for subsurface defect detection, and the film produced by the radiography method shows density changes. Very narrow defects perpendicular to the beam are hard to detect, and sometimes have to be reradiographed from a different angle to obtain a clearer image.

The main limitations with radiography are that inspection personnel have to be protected from the harmful rays and the areas have to be cleared of noninspection personnel, which means that construction has to stop.

The extent of the level or amount of radiography carried out on a piping system is defined as follows:

- *100% radiography.* This applies only to girth and miter groove welds and fabricated branch connection welds.

- *Random radiography.* This applies only to girth and miter groove welds.

- *Spot radiography.* This requires a single exposure to radiography at a specified location.

7.2.5 Ultrasonic Examination

Ultrasonic examination of components and workmanship should be performed in accordance with BPV Code, Section V, Article 5.

Ultrasonic examination uses a high-frequency (100 kHz–10 MHz) sound wave transmitted as a beam through the work piece. When the beam meets a defect it is reflected back to the source. In certain cases, ultrasonic testing can be used if radiography is not an option and it can be used to support MPE or LPE.

7.3 Testing of Piping Systems

Before a piping system can be commissioned for plant startup and after all the required NDE activities have been completed, each piping system must be tested to ensure the integrity of its pressure-containing capabilities. This test usually is a hydrostatic leak test, unless for special reasons this is not an option.

There are a number of methods for pressure and leak testing process piping systems, and the industry tests most commonly used are

- Hydrostatic testing, which uses water or another liquid under pressure.

- Pneumatic or gaseous-fluid testing, which uses air or another gas under pressure.

- A combination of pneumatic and hydrostatic testing, where low-pressure air is first used to detect leaks.

- Initial service testing, which involves a leakage inspection when the system is first put into operation.

- Vacuum testing, which uses negative pressure to check for the existence of a leak.

The owner may decided that a piping system for nonflammable, non-toxic category D fluid service may be subjected to an initial service leak test using the service fluid as the test medium. If the owner considers a hydrostatic leak test impracticable, it can be replaced by either a pneumatic test or a combined hydrostatic-pneumatic test, as long as the potential danger of stored energy of a gas is recognized and the necessary safety precautions are taken.

Where the owner of the plant considers that hydrostatic testing might damage or contaminate the internal components of the piping system or that a pneumatic test is too hazardous, a 100% radiography or ultrasonic testing of all welds in the piping system is an option. This is a costly exercise, but sometimes there is no alternative.

7.3.1 Some Limitations on the Pressure Testing of Piping Systems

If the test pressure is likely to produce a nominal pressure stress or longitudinal stress in excess of yield strength at the test temperature, then the test pressure may be reduced to the maximum pressure that will not exceed the yield strength at the test temperature. If a pressure test has to be maintained for a period of time and the test fluid in the system is subjected to thermal expansion, precautions should be taken to avoid excessive pressure.

A preliminary pneumatic test using air at no more than 170 kPa (25 psi) gauge pressure may be made prior to hydrostatic testing to locate major leaks. A leak test must be maintained for a minimum of 10 minutes or the length of time that it takes to visually inspect each pipe weld and mechanical connection for leaks.

All heat treatment required for the piping system must be completed prior to the leak test. Painting and insulation also must not be started until the piping system has successfully passed all leak tests. In the case of double-containment jacketed piping systems, the internal pressure containing piping system must successfully complete a leak test, before the external jacket is added. Painting may mask a low-pressure leak and the piping systems must be uninsulated and unjacketed so that all welds are visible for examination during the leak test. For double-containment piping systems, the external jacketed piping system is leak tested after the successful completion of the internal piping system.

7.3.2 Special Provisions for Testing

Piping subassemblies or sections of a complete piping system can be tested separately, if testing a system in its entirety is not possible or logistically difficult. The final weld, or closure weld that joins two connecting piping systems that have been successfully leak tested, need not be part of a leak test, if it is 100% circumferencially radiographed.

If a piping system fails a leak test, then it must be repaired where necessary, using the same procedures as the original joint and then subjected to a retest.

7.3.3 Preparation for Leak Test

All joints and welds on the piping system, including structural attachment welds, such as pipe supports to pressure-containing components, and bonds should be left uninsulated and exposed for examination during leak testing.

Piping systems designed for vapor or gas should be provided with additional temporary supports, if required, to support the weight of liquid during the leak test.

A piping system that has expansion joints that depend on external main anchors to restrain pressure-end loads should be tested in place. A system that has self-restrained expansion joints, which have been previously shop tested by the manufacturer, can be excluded from the system being tested. Bellows-type expansion joints must not be subjected to a leak-test pressure greater than the manufacturer's test pressure.

Items of equipment that are not subjected to a leak test can be disconnected from the piping system to be tested by isolating, using line blinds or a valve suitable for the test pressure of the system it is isolating.

7.4 Leak-Testing Methods

7.4.1 Hydrostatic Leak Test

Hydrostatic testing is the most commonly used method of leak testing and a safer method than pneumatic testing. Water is almost noncompressible and there is a limited amount of stored energy; however, gaseous test mediums used for pneumatic testing are highly compressible. This means that stored energy is contained within the piping system during the test period could be released suddenly in the event of a failure, causing local damage to personnel and equipment.

The potential damage possible during a pneumatic test is far greater than in a hydrotest; however, trapped air pockets in a hydrotest also can have a disastrous effect if there is a failure. That is why it is essential that all high points in piping systems have vents to allow air to escape as the piping system is filled with water.

Hydrotests might not be an option if all traces of water are to be avoided during operation, as in the case of cryogenic services operating at well below subzero temperatures. Also, the disposal of large volumes of water may be a problem, or conversely, water may be a rare commodity in the desert.

Leaks are detected by the visual examination of all joints for signs of water. This is not as effective as leak detecting using a gaseous medium. Gas leaks are more visible using a soapy liquid solution that covers the weld, and bubbles are detected in the event of a leak. If helium is used as the test medium, then more sophisticated equipment can be used, such as gas sniffers that detect minute leakages.

Water is not as sensitive as gases because of the surface tension. This can create a barrier, and minute leak paths actually may be protected.

According to ASME B31.3, the hydrostatic test pressure at any point in a metallic piping system should be as follows: no less than 1½ times the design pressure and, for design temperatures above the test temperature, the minimum test pressure should be calculated by the following equation, except that the value of ST/S should not exceed 6.5—

$$P_r = \frac{1.5 P S_r}{S} \qquad (7.1)$$

where

P = internal design gauge pressure.

PT = minimum test gauge pressure.

S = stress value at design temperature (Table A-1, ASME B31.3).

ST = stress value at test temperature.

If the test pressure calculated using this equation produces a nominal pressure stress or longitudinal stress that exceeds the yield strength at the test temperature, then the test pressure may be reduced to the maximum pressure that does not exceed the yield strength at the test temperature.

If it is not practical to isolate a vessel and a piping system to be leak tested in combination with a pressure vessel, then the lesser of the two test pressures should be used.

7.4.2 Pneumatic Leak Test

The fluids most commonly used for a pneumatic test are air or nitrogen; however, gas testing involves the hazard of the piping system retaining stored energy that could have a very damaging effect if there is a failure. Great care must be taken to minimize the chance of a failure during a pneumatic leak test. The test temperature is important in this regard and must be considered when the designer chooses the material of construction.

Nitrogen should not be used in a closed area if there is possibility that escaping nitrogen might displace the air in the confined space and personnel could be trapped in an environment with low oxygen levels.

ASME B31.3 quotes a pneumatic test pressure of 110% of design pressure and a pressure relief device must be installed and set so that the test pressure will exceed only the lesser of 345 kPa (50 psi) or 10% of the test pressure.

Because of the risk element during a pneumatic test, the pressure should be increased gradually, in steps; and there must be a duration between these steps when the piping system is allowed to stabilize.

7.4.3 Combination Hydrostatic-Pneumatic Leak Test

For a combined hydrotest-pneumatic test of a piping system, first, a low air pressure, of approximately 25 psig (175 kPa), is introduced to see if there are any major leaks. This low pressure is less dangerous than a full pneumatic test, and major leaks can be detected easily by the soapy water bubble test. Necessary repairs can be carried out before advancing to the hydrotest.

7.4.4 Initial Service Leak Testing

The code ASME B31.3 limits the use of this technique to category D fluid service. The pressure of the service and test fluid should be gradually increased in predetermined steps until the operating pressure is reached, and the pressure should be allowed to stabilize for a period at each step, long enough to allow equalization to take place in the piping system.

Records must be kept for each piping system or subassembly during the testing, and it must include

- Date of test.
- Identification of piping system tested.
- Test fluid.
- Test pressure.
- Certification of results by examiner.

These records need not be archived as a permanent record if certification is issued by the inspector that confirms the successful completion of the leak test.

7.5 Choice of Testing Medium

The leak-testing method selected must be compatible with the materials of construction piping-system requirements and the service fluid. This information is available at an early stage of the project, and so the test medium can be defined. As mentioned previously, a hydrostatic test is preferred over pneumatic testing because of the danger element of the latter.

7.6 Test Pack

Prior to leak testing, the test engineer responsible for the piping system must compile a test pack that contains sufficient information for the test contractor to complete a test. The test pack includes

- P&ID to show the extent of the test with boundary isolation points identified.
- All relevant isometrics.
- Test record sheet.
- Valve sheet that shows the positions of the valves (open, closed, half open).

7.7 Punch List

A punch list must be prepared prior to a leak test, this is a checklist to ensure that all the work has been completed on the piping system and the test can commence in safety. This exercise involves the following activities:

- Review of P&ID drawings to ensure that the piping system is complete.
- Piping isometrics for alignment with the P&IDs.
- Completed and torqued flanges with no missing bolts or gaskets.
- All gravity supports installed.
- Correct pipe routing.
- Correct valve type and orientation.
- Vents and drains installed to allow proper filling, venting, and draining at high and low points.
- Proper material type verified using color codes or markings, and heat numbers recorded, if required by the codes.

- Piping stress relief, weld examinations, and welding documentation completed.

- Disconnection or isolation by closed valves or testing blinds of equipment not to be tested.

- Assurance that any valves used to isolate the test boundaries are in place to protect both the testing personnel and any others who may be on site.

- Assurance that all nonboundary valves in the test system are in the open position.

- Assurance that expansion joints, if installed, have required restraints to protect against damage from the test pressure.

- Assurance that all springs have travel stops to protect against the weight of the test medium.

- Calibration of test equipment.

- Assurance that all test connections are tight.

For gas systems, additional gravity supports may be required temporarily to support the weight of the test liquid.

APPENDIX A

Listed Material

The following is a list of ASTM materials referenced in the design code ASME B31.3. A plant designed to this code is not limited to use only these materials; however, if materials outside of this list are to be used, then approval is required and submission of a detailed material data sheets that contains chemical composition and mechanical properties is a prerequisite.

The data sheets go to

> ASTM
> ASTM American Society for Testing and Materials
> 100 Barr Harbor Drive
> West Conshohocken, Pennsylvania 19428-2959
> 610 832-9500
> www.astm.org

A.1   ASTM Materials

The ASTM materials are as follows.

A.1.1   Ferrous Metals

> A.36, structural steel.
> A.47, ferritic malleable iron castings.

A.48, gray iron castings.

A.53, pipe, steel, black and hot-dipped, zinc coated, welded and seamless.

A.105, forgings, carbon steel, for piping components.

A.106, seamless carbon steel pipe for high-temperature service.

A.126, gray cast iron castings for valves, flanges, and pipe fittings.

A.134, pipe, steel, electric-fusion (arc)-welded (sizes NPS 16 and Over).

A.135, electric-resistance-welded steel pipe.

A.139, electric-fusion (arc)-welded steel pipe (NPS 4 and over).

A.167, stainless and heat-resisting chromium-nickel steel plate, sheet and strip.

A.179, seamless cold-drawn low-carbon steel heat- exchanger and condenser tubes.

A.181, forgings, carbon steel for general purpose piping.

A.182, forged or rolled alloy-steel pipe flanges, forged fittings, and valves and parts for high-temperature service.

A.197, cupola malleable iron.

A.202, pressure vessel plates, alloy steel, chromium-manganese-silicon.

A.203, pressure vessel plates, alloy steel, nickel.

A.204, pressure vessel plates, alloy steel, molybdenum.

A.216, steel castings, carbon, suitable for fusion welding for high-temperature service.

A.217, steel castings, martensitic stainless and alloy, for pressure-containing parts suitable for high-temperature service.

A.234, piping fittings of wrought carbon steel and alloy steel for moderate and elevated temperatures.

A.240, heat-resisting chromium and chromium-nickel stainless steel plate, sheet and strip for pressure vessels.

A.268, seamless and welded ferritic stainless steel tubing for general service.

A.269, seamless And Welded Austenitic Stainless Steel Tubing For General Service.

A.278, gray iron castings for pressure-containing parts for temperatures up to 650°F.

A.283, low and intermediate tensile strength carbon steel plates, shapes and bars.

A.285, pressure vessel plates, carbon steel, low- and intermediate-tensile strength.

A.299, pressure vessel plates, carbon steel, manganese-silicon.

A.302, pressure vessel plates, alloy steel, manganese-molybdenum and manganese-molybdenum-nickel.

A.312, seamless and welded austenitic stainless steel pipe.

A.333, seamless and welded steel pipe for low-temperature service.

A.334, seamless and welded carbon and alloy-steel tubes for low-temperature service.

A.335, seamless ferritic alloy steel pipe for high-temperature service.

A.350, forgings, carbon and low-alloy steel requiring notch toughness testing for piping components.

A.351, steel castings, austenitic, austenitic-ferritic (duplex) for pressure-containing parts.

A.352, steel castings, ferritic and martensitic, for pressure-containing parts suitable for low-temperature service.

A.353, pressure vessel plates, alloy steel, 9% nickel, double normalized and tempered.

A.358, electric-fusion-welded austenitic chromium-nickel alloy steel pipe for high-temperature service.

A.369, carbon steel and ferritic alloy steel forged and bored pipe for high-temperature service.

A.376, seamless austenitic steel pipe for high-temperature central-station service.

A.381, metal-arc-welded steel pipe for use with high-pressure transmission systems.

A.387, pressure vessel plates, alloy steel, chromium-molybdenum.

A.395, ferritic ductile iron pressure-retaining castings for use at elevated temperatures.

A.403, wrought austenitic stainless steel piping fittings.

A.409, welded large diameter austenitic steel pipe for corrosive or high-temperature service.

A.420, piping fittings of wrought carbon steel and alloy steel for low-temperature service.

A.426, centrifugally cast ferritic alloy steel pipe for high-temperature service.

A.451, centrifugally cast austenitic steel pipe for high-temperature service.

A.479, stainless and heat-resisting steel bars and shapes for use in boilers and other pressure vessels.

A.487, steel castings suitable for high-pressure service.

A.494, castings, nickel and nickel alloy.

A.515, pressure vessel plates, carbon steel, for intermediate- and higher-temperature service.

A.516, pressure vessel plates, carbon steel, for moderate- and lower-temperature service.

A.524, seamless carbon steel pipe for atmospheric and lower temperatures.

A.537, pressure vessel plates, heat-treated, carbon-manganese-silicon steel.

A.553, pressure vessel plates, alloy steel, quenched and tempered 8 and 9% nickel.

A.570, hot-rolled carbon steel sheet and strip, structural quality.

A.571, austenitic ductile iron castings for pressure-containing parts suitable for low-temperature service.

A.587, electric-welded low-carbon steel pipe for the chemical industry.

A.645, pressure vessel plates, 5% nickel alloy steel, specially heat treated.

A.671, electric-fusion-welded steel pipe for atmospheric and lower temperatures.

A.672, electric-fusion-welded steel pipe for high-pressure service at moderate temperatures.

A.691, carbon and alloy steel pipe, electric fusion-welded for high-pressure service at high temperatures.

A.789, seamless and welded ferritic/austenitic stainless steel tubing for general service.

A.790, seamless and welded ferritic/austenitic stainless steel pipe.

A.815, wrought ferritic, ferritic/austenitic and martensitic stainless steel fittings.

A.1.2 Nonferrous Material

B.21, naval brass rod, bar, and shapes.

B.26, aluminum-alloy sand castings.

B.42, seamless copper pipe, standard sizes.

B.43, seamless red brass pipe, standard sizes.

B.61, steam or valve bronze castings.

B.62, composition bronze or ounce metal castings.

B.68, seamless copper tube, bright annealed.

B.75, seamless copper tube.

B.88, seamless copper water tube.

B.96, copper-silicon alloy plate, sheet, strip, and rolled bar for general purposes and pressure vessels.

B.98, copper-silicon alloy rod, bar, and shapes.

B.127, nickel-copper alloy (UNS N04400) plate, sheet, and strip.

B.133, copper rod, bar, and shapes.

B.148, aluminum-bronze castings.

B.150, aluminum-bronze rod, bar, and shapes.

B.152, copper sheet, strip, plate, and rolled bar.

B.160, nickel rod and bar.

B.161, nickel seamless pipe and tube.

B.162, nickel plate, sheet, and strip.

B.164, nickel-copper alloy rod, bar, and wire.

B.165, nickel-copper alloy (UNS N04400) seamless pipe and tube.

B.166, nickel-chromium-iron alloy (UNS N06600) rod, bar, and wire.

B.167, nickel-chromium-iron alloy (UNS N06600-N06690) seamless pipe and tube.

B.168, nickel-chromium-iron alloy (UNS N06600-N06690) plate, sheet, and strip.

B.169, aluminum bronze plate, sheet, strip, and rolled bar.

B.171, copper-alloy condenser tube plates.

B.187, copper bar, bus bar, rod, and shapes.

B.209, aluminum and aluminum-alloy sheet and plate.

B.210, aluminum-alloy drawn seamless tubes.

B.211, aluminum-alloy bars, rods, and wire.

B.221, aluminum-alloy extruded bars, rods, wire, shapes, and tubes.

B.241, aluminum-alloy seamless pipe and seamless extruded tube.

B.247, aluminum-alloy die, hand, and rolled ring forgings.

B.280, seamless copper tube for air conditioning and refrigeration fluid service.

B.283, copper and copper-alloy die forgings (hot-pressed).

B.265, titanium and titanium alloy strip, sheet, and plate.

B.333, nickel-molybdenum alloy plate, sheet, and strip.

B.335, nickel-molybdenum alloy rod.

B.337, seamless and welded titanium and titanium alloy pipe.

B.345, aluminum-alloy seamless extruded tube and seamless pipe for gas and oil transmission and distribution piping systems.

B.361, factory-made wrought aluminum and aluminum alloy welding fittings.

B.366, factory-made wrought nickel and nickel-alloy welding fittings.

B.381, titanium and titanium alloy forgings.

B.407, nickel-iron-chromium alloy seamless pipe and tube.

B.409, nickel-iron-chromium alloy plate, sheet, and strip.

B.423, nickel-iron-chromium-molybdenum-copper alloy (UNS N08825 and N08221) seamless pipe and tube.

B.424, nickel-iron-chromium-molybdenum-copper alloy (UNS N08825 and N08221) plate, sheet, and strip.

B.425, nickel-iron-chromium-molybdenum-copper alloy (UNS N08825 and N08221) rod and bar.

B.435, UNS N06022, UNS N06230, and UNS R30556 plate, sheet, and strip.

B.443, nickel-chromium-molybdenum-columbium alloy (UNS N06625) plate, sheet, and strip.

B.444, nickel-chromium-molybdenum-columbium alloy (UNS N06625) seamless pipe and tube.

B.446, nickel-chromium-molybdenum-columbium alloy (UNS N06625) rod and bar.

B.462, forged or rolled UNS N08020, UNS N08024, UNS N08026, and UNS N08367 alloy pipe fittings, and valves and parts for corrosive high-temperature service.

B.463, forged or rolled UNS N08020, UNS N08026, and UNS N08024 alloy plate, sheet, and strip.

B.464, welded chromium-nickel-iron-molybdenum-copper-columbium stabilized alloy (UNS N08020) pipe.

B.466, seamless copper-nickel pipe and tube.

B.467, welded copper-nickel pipe.

B.491, aluminum and aluminum alloy extruded round tubes for general-purpose applications.

B.493, zirconium and zirconium alloy forgings.

B.514, welded nickel-iron-chromium alloy pipe.

B.517, welded nickel-chromium-iron alloy (UNS N06600, UNS N06603, UNS N06025, and UNS N06045) pipe.

B.523, seamless and welded zirconium and zirconium alloy tubes.

B.547, aluminum and aluminum-alloy formed and arc-welded round tube.

B.550, zirconium and zirconium alloy bar and wire.

B.551, zirconium and zirconium alloy strip, sheet, and plate.

B.564, nickel alloy forgings.

B.574, low-carbon nickel-molybdenum-chromium alloy rod.

B.575, low-carbon nickel-molybdenum-chromium alloy plate, sheet, and strip.

B.581, nickel-chromium-iron-molybdenum-copper alloy rod.

B.582, nickel-chromium-iron-molybdenum-copper alloy plate, sheet, and strip.

B.584, copper alloy sand castings for general applications.

B.619, welded nickel and nickel-cobalt alloy pipe.

B.620, nickel-iron-chromium-molybdenum alloy (UNS N08320) plate, sheet, and strip.

B.621, nickel-iron-chromium-molybdenum alloy (UNS N08320) rod.

B.622, seamless nickel and nickel-cobalt alloy pipe and tube.

B.625, nickel alloy plate and sheet.

B.649, Ni-Fe-Cr-Mo-Cu low-carbon alloy (UNS N08904) and Ni-Fe-Cr-Mo-Cu-N low-carbon alloy (UNS N08925, UNS N08031, and UNS N08926) bar and wire.

B.658, zirconium and zirconium alloy seamless and welded pipe.

B.675, UNS N08366 and UNS N08367 welded pipe.

B.688, chromium-nickel-molybdenum-iron (UNS N08366 and UNS N08367) plate, sheet, and strip.

B.690, iron-nickel-chromium-molybdenum alloys (UNS N08366 and UNS N08367) seamless pipe and tube.

B.705, nickel-alloy (UNS N06625 and N08825) welded pipe.

B.725, welded nickel (UNS N02200/UNS N02201) and nickel-copper alloy (UNS N04400) pipe.

B.729, seamless UNS N08020, UNS N08026, UNS N08024 nickel-alloy pipe and tube.

B.804, UNS N08367 welded pipe.

E.112, methods for determining average grain size.

A.2 American Petroleum Institute

For information on 5L, line pipe, the American Petroleum Institute (API) can be reached at

>API American Petroleum Institute
>Publications and Distribution Section
>1220 L Street, NW
>Washington, DC 20005-4070
>202 682-8375
>www.api.org

APPENDIX B

General Engineering Data

TRIM STANDARD MATERIALS

OMB STANDARD TRIM DEFINITIONS

API Trim No	Nominal Trim	OMB descr.	Stem	Disc/Wedge	Seat	Min Hardness (Brinell)
1	F6	F6	410 (13Cr)	F6 (13Cr)	410 (13Cr)	250
2	304	304	304 (18Cr-8Ni)	304 (18Cr-8Ni)	304 (18Cr-8Ni)	not specified
3	-	-	(25Cr-20Ni)	310 (25Cr-20Ni)	310 (25Cr-20Ni)	not specified
4	-	-	410 (13Cr)	F6 (13Cr)	F6 (13Cr)	750
5	Hardfaced	F6HF	410 (13Cr)	F6 + St Gr6 (CoCr Alloy)	410 + St Gr6 (CoCr Alloy)	350
5A	-	-	410 (13Cr)	F6+Hardf. NiCr Alloy	410+Hardf. NiCr Alloy	350
6	-	-	410 (13Cr)	F6 (13CR)	Monel® (NiCu Alloy)	250/175
7	-	-	410 (13Cr)	F6 (13Cr)	F6 (13Cr)	250/750
8	F6 and Hardfaced	F6HFS	410 (13Cr)	F6 (13Cr)	410 + St Gr6 (CoCr Alloy)	250/350
8A	-	-	410 (13Cr)	F6 (13Cr)	410 Hardf. Nicr Alloy	250/350
9	Monel	Monel	Monel® (NiCu Alloy)	Monel® (NiCu Alloy)	Monel® (NiCu Alloy)	not specified
10	316	316	316 (18Cr-8Ni-Mo)	316 (18Cr-8Ni-Mo)	316 (18Cr-8Ni-Mo)	not specified
11	Monel and Hardfaced	MonelHFS	Monel® (NiCu Alloy)	Monel® (NiCu Alloy)	Monel® St Gr6	350
11A	-	-	Monel® (NiCu Alloy)	Monel® (NiCu Alloy)	Monel® Hardif. NiCrA	350
12	316 and Hardfaced	316HFS	316 (18Cr-8Ni-Mo)	316 (18Cr-8Ni-Mo)	316 + St. Gr6	350
12A	-	-	316 (18Cr-8Ni-Mo)	316 (18Cr-8Ni-Mo)	316 Hardf. NiCr Alloy	350
13	Alloy 20	Alloy 20	Alloy 20 (19Cr-29Ni)	Alloy 20 (19Cr-29Ni)	Alloy 20 (19Cr-29Ni)	not specified
14	Alloy 20 and Hardfaced	Alloy 20HFS	Alloy 20 (19Cr-29Ni)	Alloy 20 (19Cr-29Ni)	Alloy 20 St Gr6	350
14A	-	-	Alloy 20 (19Cr-29Ni)	Alloy 20 (19Cr-29Ni)	Alloy 20 Hardf. NiCr Alloy	350
15	Hardfaced (304)	304-HF	304 (18Cr-8Ni)	304 + St Gr6	304 + St Gr6	350
16	Hardfaced (316)	316-HF	316 HF (18Cr-8Ni-Mo)	316 + St Gr6	316 + St Gr6	350
17	Hardfaced (347)	347-HF	347 HF (18Cr-10Ni-Cb)	347 + St Gr6	347 + St Gr6	350
18	Hardfaced Alloy 20 HF	Alloy 20 HF	Alloy 20 (19Cr-29Ni)	Alloy 20 + St Gr6	Alloy 20 + St Gr6	350
n/a	Alloy 625	Alloy 625	Alloy 625	Alloy 625	Alloy 625	

OMB TRIM MATERIAL

OMB	UNS	TYPE	Grade (forged)	ASTM wrought	DIN	DIN W NO.
F6	UNS S41000	13Cr	ASTM A182 F6a	A276-410	DIN X12Cr13	1,4006
304	UNS S30400	18-8 Cr-Ni	ASTM A182 F304	A276-304	DIN X5CrNi 18 10	1,4301
316	UNS S31600	18-8 Cr-Ni (18-10-2)	ASTM A182 F316	A276-316	DIN X5CrNiMo 18 10	1,4401
321	UNS S32100	18 Cr-10 Ni-Ti	ASTM A182 F321	A276-321	DIN X6CrNiTi 18 10	1.4541
347	UNS S34700	18 Cr-10 Ni-Cb	ASTM A182 F347	A276-347	DIN X6CrNiNb18 10	1.4550
MONEL(R)	UNS N04400	67Ni-30Cu	ASTM B564-N04400	B164-N04400	DIN 17743	2.4360
ALLOY 20	UNS N08020	28Ni-19Cr-Cu-Mo	ASTM A182-F20	ASTM B473	DIN 14500	2.4660
ALLOY 625	UNS N06625	60Ni-22Cr-9Mo-3.5Cb	ASTM B564-N06625	ASTM B564-N06625	DIN 17361	2.4865
C276	UNS N10276	54Ni-15Cr-16Mo	ASTM B564-N10276	ASTM B574-N10276	DIN NiMo 16 Cr 15 W	2,4819
St. Gr6	UNS R30006	Co Cr-A	AMS 5894		Stellite(R) Gr6	

Figure B–1 API Standard 600 trim materials. (Printed with the kind permission of Valvosider, srl, Italy.)

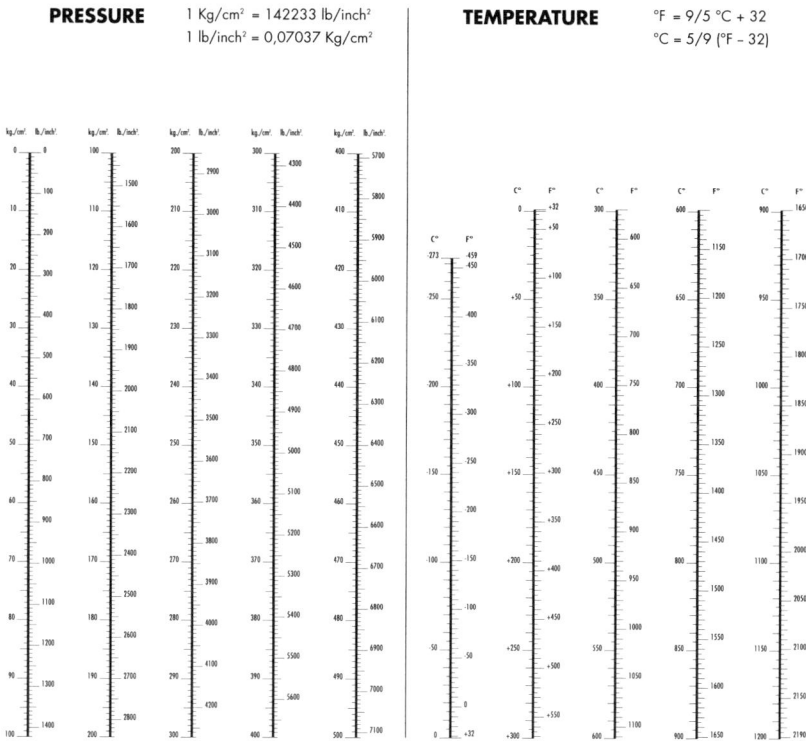

Figure B-2 *Conversion chart—pressure and temperature. (Printed with the kind permission of OMB Valves, spa, Italy.)*

222   Appendix B—General Engineering Data

CORROSIVE MEDIA	Carbon Steel	Stainless Steel 304	Stainless Steel 316	Inconel	Monel
Acetate Solvents, Crude	D	A	A	A	B
Acetate Solvents, Pure	C	A	A	A	A
Acetic Acid, 95%	D	B	A	A	A
Acetic Anhydride, Boiling	D	B	A	A	A
Acetone	B	A	A	A	A
Alcohols	B	A	A	A	A
Amines	B	A	A	A	A
Ammonia, Anhydrous	B	A	A	A	A
Ammonium Hydroxide, Hot	B	A	A	A	D
Ammonium Nitrate	B	A	A	A	C
Aniline Hydrochloride	D	D	C	B	B
Antimony Trichloride	D	D	C	B	B
Asphalt	B	A	A	A	A
Barium Chloride, 5%	C	A	A	A	A
Barium Hydroxide	C	A	A	A	A
Barium Nitrate	C	A	A	B	C
Benzene, Hot	B	A	A	A	A
Benzoic Acid	B	A	A	A	A
Blood	D	A	A	A	A
Bromine, Dry Gas	D	A	A	B	A
Bromine, Moist Gas	D	D	D	D	C
Buttermilk	D	A	A	A	A
Calcium Bisulfite, Hot	D	C	B	D	D
Calcium Chloride, Dilute	C	B	A	A	A
Calcium Hydroxide, 20%, Boiling	D	A	A	A	A
Calcium Hydrochloride, < 2%	C	C	B	B	C
Carbolic Acid, 90%	C	A	A	A	B
Carbon Dioxide, Dry	C	A	A	A	A
Carbon Disulphide	B	A	A	A	B
Chloroacetic Acid	D	D	C	B	B
Chloric Acid	D	D	C	C	C
Chlorinated Water, Sat.	D	D	C	C	C
Chlorine, Dry Gas	B	B	B	A	A
Chlorine, Moist Gas	D	D	C	D	C
Citric Acid, Dilute	D	A	A	A	A
Citric Acid, Hot, Conc.	D	C	B	B	B
Creosote, Hot	B	A	A	A	A
Cupric Chloride, 5%	D	D	C	D	D
Ethyl Chloride	A	A	A	A	A
Ethylene Glycol	A	A	A	A	A
Ferric Chloride < 1%	D	C	B	B	C
Ferric Nitrate, 5%	D	B	A	C	D
Ferric Sulfate, 5%	D	B	A	B	C
Ferrous Sulfate, 10%	C	A	A	B	A
Flourine, Dry Gas	C	C	B	A	A
Flourine, Moist Gas	D	D	D	B	A
Freon, Wet	C	C	C	B	A
Fuel Oil, 140°F	A	A	A	A	B

CORROSIVE MEDIA	Carbon Steel	Stainless Steel 304	Stainless Steel 316	Inconel	Monel
Furfural	B	B	B	B	B
Gasoline Sour	B	A	A	C	C
Gasoline Refined	A	A	B	A	A
Gelatine	D	B	A	A	A
Glucose	B	A	A	A	A
Glycerine	B	A	A	A	A
Hydrofluoric Acid, Boiling	D	D	D	D	B
Hydrofluosilicic Acid	D	D	C	B	A
Hydrogen Chloride, Dry	B	D	C	B	A
Hydrogen Chloride, Moist	D	D	D	D	C
Hydrogen Fluoride, Dry	C	D	C	A	A
Hydrogen Peroxide, Boiling	D	C	B	B	B
Hydrogen Sulfide, Dry	B	A	A	A	A
Hydrogen Sulfide, Moist	C	B	A	A	B
Iodine, Dry	D	D	B	A	A
Kerosene	A	A	A	A	A
Lactic Acid, 5%	D	B	A	A	B
Lactic Acid, 10%	D	B	A	A	B
Lactic Acid, Boiling, 5%	D	C	B	B	C
Lactic Acid, Boiling, 10%	D	D	B	B	C
Lead Acetate, Hot	D	A	A	B	B
Magnesium Chloride, Hot, 5%	D	C	B	A	A
Magnesium Hydroxide	B	A	A	A	A
Magnesium Sulfate	B	A	A	B	A
Magnesium Sulfate, Boiling	C	A	A	C	A
Mercury	B	A	A	A	B
Mercuric Chloride, < 2%	D	D	D	D	D
Mercuric Cyanide	D	B	B	B	D
Methyl Chloride, Dry	D	B	B	A	A
Milk	D	A	A	A	B
Molasses	B	A	A	A	A
Naptha	B	A	A	A	A
Nickel Chloride	D	C	B	B	B
Nickel Sulfate, Boiling	D	C	C	B	A
Nitric Acid, 20%	D	A	A	B	D
Nitric Acid, Boiling, Conc.	D	D	D	D	D
Nitrobenzene	D	B	A	B	B
Nitrous Acid	D	B	B	B	C
Oils - Miner.	B	A	A	C	B
Oxalic Acid, Boiling, 10%	C	A	A	A	A
Oxalic Acid, Boiling, 50%	D	D	C	B	B
Oxygen	B	A	A	A	A
Picric Acid	C	A	A	D	D
Potassium Bromide	D	C	B	A	A
Potassium Carbonate	B	A	A	A	A
Potassium Chlorate	B	A	A	A	B
Potassium Chloride	D	A	A	A	A
Potassium Chloride, Hot	D	C	B	B	A

CORROSIVE MEDIA	Carbon Steel	Stainless Steel 304	Stainless Steel 316	Inconel	Monel
Potassium Cyanide	B	B	B	B	B
Potassium Sulfate, Dil	B	A	A	A	A
Propane, Liquid & Gas	B	A	A	A	A
Pyrogallic Acid	B	A	A	B	A
Rosin, Molten	D	A	A	A	A
Salicylic Acid	D	B	B	B	B
Silver Bromide	D	B	A	C	B
Silver Chloride	D	D	D	C	B
Silver Nitrate	D	A	A	A	C
Sodium Acetate	C	A	A	A	A
Sodium Bisulfate	D	B	B	B	A
Sodium Bromide, Dil.	D	B	B	B	A
Sodium Cyanide	B	B	B	B	A
Sodium Fluoride, 5%	D	B	A	B	A
Sodium Hydroxide, 50%	B	A	A	A	A
Sodium Hyposulfite	D	B	A	B	A
Sodium Nitrate	B	B	A	A	B
Sodium Perborate	C	A	A	A	B
Sodium Peroxide	C	A	A	A	B
Sodium Phosphate, Tribasic	C	A	A	A	A
Sodium Silicate	B	A	A	B	A
Sodium Thiosulfate	D	B	A	B	B
Stannous Chloride, Sat.	D	D	B	B	B
Steam, 212°F	A	A	A	A	A
Steam, 600°F	C	A	A	A	A
Sulfite Liquors	D	C	B	D	D
Sulfur Chloride	D	C	D	B	B
Sulfur Dioxide, Moist	D	B	A	D	D
Sulfuric Acid, Conc.	B	B	B	B	D
Sulfurous Acid, Sat.	D	B	B	D	D
Tannic Acid, 10%	D	A	A	B	A
Tar, Hot	B	A	A	A	B
Tartaric Acid, 120°F	D	B	A	A	A
Toluene	A	A	A	A	A
Trichlorethylene	B	A	A	A	A
Turpentine	B	A	A	A	A
Varnish, Hot	C	A	A	A	A
Vegetable Oils	B	A	A	A	B
Vinegar	D	A	A	A	A
Water, Acid Mine	D	A	A	A	C
Water, Boiler Feed	B	A	A	A	A
Water, Distilled	D	A	A	A	A
Water, Salt Sea	D	C	B	B	A
Whiskey, Boiling	D	A	A	A	C
Wine	D	A	A	A	A
Xylene, Boiling	D	A	A	A	A
Zinc Chloride, 5%	D	C	B	B	B
Zinc Sulfate, Boiling	D	B	A	B	A

A = Substantial resistance - Preferred material of construction.
B = Moderate resistance - Satisfactory for use under most conditions.
C = Questionable resistance - Use with caution.
D = Inadequate resistance - Not recommended.

OMB doesn't assume any responsibility from the use of a.m. data which are purely theoretical. The user must verify the best conditions of use.

Figure B–3   *Material equivalents. (Printed with the kind permission of OMB Valves, spa. Italy.)*

Appendix B—General Engineering Data 223

Material Group	Common Name	Nominal Type	UNS	Forging Spec.	Casting Spec. Equivalent	DIN	DIN W. No	Application Notes
Carbon Steel	CS	C-Mn-Fe	K03504	A105N	A216-WCB	C22.8 DIN 17243	1.0460	General non-corrosive service from -20F(-29C) to 800F(427C)
Low Temperature Carbon Steel	LTCS	C-Mn-Fe	K03011	A350-LF2	A352-LCA A352-LCB A352-LCC	TSTE 355 DIN 18103	1.0566	General non-corrosive service from -50F (-46C) to 650F(340C), LF2 to 800F(427C).
Low Temperature Alloy Steel	Nickel Steel	3.1/2Ni	K32025	A350-LF3	A352-LC3	10Ni14	1.5637	-150F(-101C) to 650F(340C)
Low Alloy Steel	Moly Steel	C-1/2Mo	K12822	A182-F1	A217-WC1	15M03	1.5415	Up to 875F (468C)
	Alloy Steel Chrome Moly	1.1/4Cr-1/2Mo	K11572	A182-F11 cl2	A217-WC6	13CRM044	1.7335	Up to 1100F (593C)
		2.1/4Cr-1Mo	K21590	A182-F22 cl3	A217-WC9	10CRM0910	1.7380	Up to 1100F(593C), HP steam
		5Cr-1/2Mo	K41545	A182-F5	A217-C5	12CRM0195	1.7362	High temp refinery service
		9Cr-1Mo	K90941	A182-F9	A217-C12	X 12 CrMo 9 1	1.7386	High temp erosive refinery service
		9Cr-1Mo-V		A182-F91	A217-C12A	X 10 CrMoVNb 9 1	1.4903	High pressure steam
Stainless Steel	Austenitic S.Steel 300 series S.Steel	304 : 18Cr-8Ni	S30400	A182-F304	A351-CF8	DIN X5CrNi 18 9	1.4301	0.04% min. carbon for temp.>1000F(538C)
		304L : 18Cr-8Ni	S30403	A182-F304L	A351-CF3	X 2 CrNi 19 11	1.4306	Up to 800F(427C)
		304H :	S30409	A182-F304H		n/a	n/a	
		316 : 16Cr-12Ni-2Mo	S31600	A182-F316	A351-CF8M	DIN X5CrNiMo 18 10	1.4401	0.04% min. carbon for temp.>1000F(538C)
		316L : 16Cr-12Ni-2Mo	S31603	A182-F316L	A351-CF3M	X 5 CrNiMo 17 12 2	1.4404	Up to 800F(427C)
		316H :	S31609	A182-F316H		n/a	n/a	
		316Ti :	S31635	A182-F316Ti		X 6 CrNiMoTi 17 12 2	1.4571	
		321: 18Cr-10Ni-Ti	S32100	A182-F321		X 6 CrNiTi 18 10	1.4541	0.04% min. carbon (grade F321H) and heat treat at 2000F(1100C) for service temps.>1000F(538C)
		321H	S32109	A182-F321H		n/a	n/a	
		347: 18Cr-10Ni-Cb(Nb)	S34700	A182-F347	A351-CF8C	DIN 8556	1.4550	0.04% min. carbon (grade F347H) and heat treat at 2000F(1100C) for service temps.>1000F(538C)
		347H	S34709	A182-F347H		n/a	n/a	
		317L	S31703	A182-F317L	A351-CG3M	X2CrNiMo18-16-4	1.4438	
	Alloy 20	28Ni-19Cr-Cu-Mo	N08020	A182-F20	A351-CN7M	DIN 1.4500	2.4660	service to 600F(316C)
	Duplex 2205	22Cr-5Ni-3Mo-N	S31803 S32205	A182-F51	A890-J92205	X2CrNiMoN22-5-3 DIN 10088-1 (95)	1.4462	service to 600F(316C) -The original S31803 UNS designation has been supplemented by S32205 which has higher minimum N, Cr, and Mo.
	Super Duplex 2507	25Cr-7Ni-4Mo-N	S32750	A182-F53	A351-CD4MCu A890 5A	X2CrNiMoN25-7-4 DIN 10088-1 (95)	1.4501	service to 600F(316C)
	Super Austenitic 6Mo	20Cr-18Ni-6Mo	S31254	A182-F44	A351-CK3MCuN	X1CrNiMoCuN20-18-7 DIN 10088-1 (95)	1.4547	service to 600F(316C)
Nickel-Iron Alloy	Incoloy 800	33Ni-42Fe-21Cr	N08800	B564-N08800		X10NiCrAlTi32-20	1.4876	service to 1000F(538C)
	Incoloy 825	42Ni-21.5Cr-3Mo-2.3Cu	N08825	B564-N08825	A494-CU5MCuC	DIN 17744	2.4858	service to 600F(316C) for N02200, 1200F(648C) for N02201
Nickel	Nickel	99/95Ni	N02200	B160-N02200 (bar)	A494-CZ-100	NW2200	1.7740	
Nickel-Copper	Monel 400	67Ni-30Cu	N04400	B564-N04400	A494-M35-1	DIN 17730	2.4360	
	Monel 500		N05500	B564-N05500			2.4375	
Nickel-Alloy	904L		N08904	904L	n/a	Z2 NCDU 25-20	1.4539	
Nickel Superalloys	Inconel 600	72Ni-15Cr-8Fe	N06600	B564-N06600	A494-CY40	DIN 17742	2.4816	
	Inconel 625	60Ni-22Cr-9Mo-3.5Cb	N06625	B564-N06625*	A494-CW-6MC		2.4856	*Difficult to forge in close dye
	Hastelloy C-276	54Ni-15Cr-16Mo	N10276	B564-N10276*	A494-CW-2M	NiMo 16 Cr 15 W	2.4819	*Difficult to forge in close dye
Titanium	Titanium	98Ti	R50400	B381-Gr2	B367-C2	Ti 2	3.7035	

Figure B–4 *Pressure and temperature ratings—material groups 1.1 and 1.9. (Printed with the kind permission of OMB Valves, spa, Italy.)*

CLASS 800

SERVICE TEMPER.	A105[1] A350-LF2[1]	A182[2] F11	A182[2] F22	A182 F5	A182 F9	A182 F304	A182 F316	A182 F304L	A182 F347H	SERVICE TEMPER.	A105[1] A350-LF2[1]	A182[2] F11	A182[2] F22	A182 F5	A182 F9	A182 F304	A182 F316	A182 F304L	A182 F347H
°F	psi	psi	psi	psi	psi	psi	psi	psi	psi	°C	bar	bar	bar	bar	bar	bar	bar	bar	bar
-20 to 100	1975	2000	2000	2000	2000	1920	1920	1600	1920	-29 to 38	136.2	137.9	137.9	137.9	137.9	132.4	132.4	110.3	132.4
200	1800	1900	1910	2000	2000	1600	1655	1350	1695	93.5	124.1	131.0	131.7	137.9	137.9	110.3	114.1	93.1	116.9
300	1750	1795	1805	1940	1940	1410	1495	1210	1570	149	120.7	123.8	124.5	133.8	133.8	97.2	103.1	83.4	108.3
400	1690	1755	1730	1880	1880	1255	1370	1100	1480	204.5	116.6	121.0	119.3	129.7	129.7	86.5	94.5	75.9	102.1
500	1595	1710	1705	1775	1775	1165	1275	1020	1380	260	110.0	117.9	117.6	122.4	122.4	80.3	87.9	70.3	95.2
600	1460	1615	1615	1615	1615	1105	1205	960	1310	315.5	100.7	113.4	113.4	113.4	113.4	76.2	83.1	66.2	90.3
650	1430	1570	1570	1570	1570	1090	1185	935	1280	343.5	98.6	108.3	108.3	108.3	108.3	75.2	81.7	64.5	88.3
700	1420	1515	1515	1515	1515	1075	1150	915	1250	371	97.9	104.5	104.5	104.5	104.5	74.1	79.3	63.1	86.2
750	1345	1420	1420	1420	1420	1060	1130	895	1230	399	92.7	97.9	97.9	97.9	97.9	73.1	77.9	61.7	84.8
800	1100	1355	1355	1325	1355	1050	1105	875	1215	426.5	75.9	93.4	93.4	91.4	93.4	72.4	76.2	60.3	83.8
850	715	1300	1300	1170	1300	1035	1080	860	1185	454.5	49.3	89.7	89.7	80.7	89.7	71.4	74.5	59.3	81.7
900	460	1200	1200	940	1200	1025	1050		1150	482	31.7	82.8	82.8	64.8	82.8	70.7	72.4		79.3
950	275	1005	1005	695	985	1000	1030		1030	510	19	69.3	69.3	47.9	67.9	69.0	71.0		71.0
1000	140	595	715	510	780	860	970		970	538	9.7	41.0	49.3	35.2	53.8	59.3	66.9		66.9
1050		365	530	375	505	825	960		960	565.5		25.2	36.6	25.9	34.8	56.9	66.2		66.2
1100		255	300	275	300	685	860		860	593.5		17.6	20.7	19.0	20.7	47.2	59.3		59.3
1150		140	275	185	200	520	735		735	621		9.7	19.0	12.8	13.8	35.9	50.7		50.7
1200		95	145	120	140	415	550		460	649		6.6	10.0	8.3	9.6	28.6	37.9		31.7
1250						295	485		330	676.5						20.3	33.4		22.8
1300						218	365		250	704.5						15.0	25.2		17.2
1350						165	275		180	732.5						11.4	19.0		12.4
1400						130	200		140	760.5						9.0	13.8		9.6
1450						95	155		110	788.5						6.6	10.7		7.6
1500						65	110		95	815.5						4.5	7.6		6.6

Figure B–5 *Pressure and temperature ratings—material groups 1.10 and 2.2. (Printed with kind permission of OMB Valves, spa, Italy.)*

IMPERIAL UNITS - psig / °F - from ASME B16.34

ASTM A105 - ASTM A350 LF2-ASTM A216 WCB • ASME B16.34 GROUP 1.1

°F	150# (PN20)	300# (PN50)	600# (PN100)	800# (PN140)	1500# (PN250)	2500# (PN420)	4500# (PN760)
-20	285	740	1480	1975	3705	6170	11110
100	285	740	1480	1975	3705	6170	11110
200	260	675	1350	1800	3375	5625	10120
300	230	655	1315	1750	3280	5470	9845
400	200	635	1270	1690	3170	5280	9505
500	170	600	1200	1595	2995	4990	8980
600	140	550	1095	1460	2735	4560	8210
650	125	535	1075	1430	2685	4475	8055
700	110	535	1065	1420	2665	4440	7990
750	95	505	1010	1345	2520	4200	7560
800	80	410	825	1100	2060	3430	6170

METRIC UNITS - °C / barg - values interpolated from ASME B16.34

ASTM A105 - ASTM A350 LF2-ASTM A216 WCB • ASME B16.34 GROUP 1.1

°C	150# (PN20)	300# (PN50)	600# (PN100)	800# (PN140)	1500# (PN250)	2500# (PN420)	4500# (PN760)
-29	19.7	51.0	102.1	136.2	255.5	425.5	766.2
0	19.7	51.0	102.1	136.2	255.5	425.5	766.2
50	19.3	50.0	100.1	133.6	250.5	417.2	751.2
100	17.7	46.4	92.8	123.7	232.0	386.6	695.7
150	15.8	45.1	90.6	120.6	226.1	377.0	678.5
200	14.0	43.9	87.8	116.9	219.2	365.2	657.4
250	12.1	41.8	83.6	111.1	208.7	347.6	625.7
300	10.2	38.9	77.5	103.3	193.6	322.8	581.1
350	8.4	36.9	74.0	98.5	184.8	308.0	554.4
375	7.4	36.6	72.9	97.2	182.4	303.9	546.9
400	6.5	34.6	69.1	92.1	172.5	287.5	517.5
427	5.5	28.3	56.9	75.9	142.1	236.6	425.5

Notes: 1MPa = 10 bar; 100 kPa = 1bar

IMPERIAL UNITS - psig / °F

ASTM A182 F11CL 2 - ASTM A217 WC6 • ASME B16.34 GROUP 1.9

°F	150# (PN20)	300# (PN50)	600# (PN100)	800# (PN140)	1500# (PN250)	2500# (PN420)	4500# (PN760)
-20	290	750	1500	2000	3750	6250	11250
100	290	750	1500	2000	3750	6250	11250
200	260	750	1500	2000	3750	6250	11250
300	230	720	1445	1925	3610	6015	10830
400	200	695	1385	1850	3465	5775	10400
500	170	665	1330	1775	3325	5540	9965
600	140	605	1210	1615	3025	5040	9070
650	125	590	1175	1570	2940	4905	8825
700	110	570	1135	1515	2840	4730	8515
750	95	530	1065	1420	2660	4430	7970
800	80	510	1015	1355	2540	4230	7610
850	65	485	975	1300	2435	4060	7305
900	50	450	900	1200	2245	3745	6740
950	35	320	640	850	1595	2655	4785
1000	20	215	430	575	1080	1800	3240
1050	20	145	290	385	720	1200	2160
1100	20	95	190	255	480	800	1440
1150	20	60	125	165	310	515	925
1200	15	40	75	100	190	315	565

METRIC UNITS - °C / barg

ASTM A182 F11 CL 2 - ASTM A217 WC6 • ASME B16.34 GROUP 1.9

°C	150# (PN20)	300# (PN50)	600# (PN100)	800# (PN140)	1500# (PN250)	2500# (PN420)	4500# (PN760)
-29	20.0	51.7	103.4	137.9	258.6	431.0	775.9
0	20.0	51.7	103.4	137.9	258.6	431.0	775.9
50	19.5	51.7	103.4	137.9	258.6	431.0	775.9
100	17.7	51.5	103.0	137.3	257.5	429.1	772.4
150	15.8	49.6	99.6	132.7	248.8	414.5	746.3
200	14.0	48.1	95.8	128.0	239.8	399.6	719.6
250	12.1	46.2	92.4	123.3	231.1	385.0	692.6
300	10.2	42.9	85.8	114.5	214.4	357.2	642.8
350	8.4	40.4	80.4	107.4	201.1	335.4	603.5
375	7.4	38.9	77.6	103.6	194.1	323.3	582.0
400	6.5	36.5	73.3	97.8	183.1	305.0	548.7
425	5.6	35.3	70.2	93.7	175.7	292.6	526.3
450	4.6	33.7	67.7	90.3	169.1	281.9	507.2
475	3.7	31.7	63.4	84.6	158.2	263.9	475.0
500	2.8	25.3	50.6	67.3	126.1	210.2	378.5
525	1.9	18.2	36.3	48.4	90.8	151.3	272.5
550	1.4	12.7	25.4	33.9	63.6	105.9	190.7
575	1.4	8.8	17.7	23.5	44.0	73.4	132.1
600	1.4	6.0	12.0	16.1	30.3	50.5	90.8
625	1.3	3.9	8.1	10.8	20.2	33.6	60.3
649	1.0	2.8	5.2	6.9	13.1	21.7	39.0

Notes: 1MPa = 10 bar; 100 kPa = 1bar

Figure B–6 *Conversion chart dimensions—U.S. customary units to metric units. (Printed with kind permission of OMB Valves, spa, Italy.)*

IMPERIAL UNITS - psig / °F

ASTM A182 F22 CL 3 - ASTM A217 WC9 • ASME B16.34 GROUP 1.10

°F	150# (PN20)	300# (PN50)	600# (PN100)	800# (PN140)	1500# (PN250)	2500# (PN420)	4500# (PN760)
-20	290	750	1500	2000	3750	6250	11250
100	290	750	1500	2000	3750	6250	11250
200	260	750	1500	2000	3750	6250	11250
300	230	730	1455	1940	3640	6070	10923
400	200	705	1410	1880	3530	5880	10585
500	170	665	1330	1775	3325	5540	9965
600	140	605	1210	1615	3025	5040	9070
650	125	590	1175	1570	2940	4905	8825
700	110	570	1135	1515	2840	4730	8515
750	95	530	1065	1420	2660	4430	7970
800	80	510	1015	1355	2540	4230	7610
850	65	485	975	1300	2435	4060	7305
900	50	450	900	1200	2245	3745	6740
950	35	375	755	1005	1885	3145	5665
1000	20	260	520	695	1305	2170	3910
1050	20	175	350	465	875	1455	2625
1100	20	110	220	295	550	915	1645
1150	20	70	135	180	345	570	1030
1200	20	40	80	110	205	345	615

METRIC UNITS - °C / barg

ASTM A182 F22 CL 3 - ASTM A217 WC9 • ASME B16.34 GROUP 1.10

°C	150# (PN20)	300# (PN50)	600# (PN100)	800# (PN140)	1500# (PN250)	2500# (PN420)	4500# (PN760)
-29	20.0	51.7	103.4	137.9	258.6	431.0	775.9
0	20.0	51.7	103.4	137.9	258.6	431.0	775.9
50	19.5	51.7	103.4	137.9	258.6	431.0	775.9
100	17.7	51.6	100.1	107.1	057.7	109.5	770.0
150	15.8	50.3	100.3	133.7	250.9	418.4	753.0
200	14.0	48.8	97.5	130.0	244.1	406.6	731.9
250	12.1	46.3	92.7	123.7	231.8	386.2	694.7
300	10.2	42.9	85.8	114.5	214.4	357.2	642.8
350	8.4	40.4	80.4	107.4	201.1	335.4	603.5
375	7.4	38.9	77.6	103.6	194.1	323.3	582.0
400	6.5	36.5	73.3	97.8	183.1	305.0	548.7
425	5.6	35.3	70.2	93.7	175.7	292.6	526.3
450	4.6	33.7	67.7	90.3	169.1	281.9	507.2
475	3.7	31.7	63.4	84.6	158.2	263.9	475.0
500	2.8	27.7	55.7	74.2	138.9	231.8	417.4
525	1.9	21.6	43.3	57.8	108.4	180.6	325.3
550	1.4	15.4	30.7	41.0	77.0	128.0	230.7
575	1.4	10.5	21.1	28.1	52.7	87.7	158.1
600	1.4	6.9	13.8	18.4	34.5	57.4	103.3
625	1.4	4.5	8.8	11.7	22.4	37.1	67.0
649	1.4	2.8	5.5	7.6	14.1	23.8	42.4

Notes: 1 MPa = 10 bar; 100 kPa = 1 bar

IMPERIAL UNITS - psig / °F

ASTM A182 F316 - ASTM A351 CF8M • ASME B16.34 GROUP 2.2

°F	150# (PN20)	300# (PN50)	600# (PN100)	800# (PN140)	1500# (PN250)	2500# (PN420)	PN420 (BAR)	4500# (PN760)
-20	275	720	1440	1920	3600	6000	413.8	10800
100	275	720	1440	1920	3600	6000	413.8	10800
200	235	620	1240	1655	3095	5160	355.9	9290
300	215	560	1120	1495	2795	4660	321.4	8390
400	195	515	1025	1370	2570	4280	295.2	7705
500	170	480	955	1275	2390	3980	274.5	7165
600	140	450	900	1205	2255	3760	259.3	6770
650	125	445	890	1185	2220	3700	255.2	6660
700	110	430	870	1160	2170	3620	249.7	6515
750	95	425	855	1140	2135	3560	245.5	6410
800	80	420	845	1125	2110	3520	242.8	6335
850	65	420	835	1115	2090	3480	240.0	6265
900	50	415	830	1105	2075	3460	238.6	6230
950	35	385	775	1030	1930	3220	222.1	5795
1000	20	350	700	935	1750	2915	201.0	5245
1050	20	345	685	915	1720	2865	197.6	5155
1100	20	305	610	815	1525	2545	175.5	4575
1150	20	235	475	630	1185	1970	135.9	3550
1200	20	185	370	495	925	1545	106.6	2775
1250	20	145	295	390	735	1230	84.8	2210
1300	20	115	235	310	585	970	66.9	1750
1350	20	95	190	255	480	800	55.2	1440
1400	20	75	150	200	380	630	43.4	1130
1450	20	60	115	155	290	485	33.4	875
1500	20	40	85	110	205	345	23.8	620

METRIC UNITS - °C / barg

ASTM A182 F316 - ASTM A351 CF8M • ASME B16.34 GROUP 2.2

°C	150# (PN20)	300# (PN50)	600# (PN100)	800# (PN140)	1500# (PN250)	2500# (PN420)	4500# (PN760)
-29	19.0	49.7	99.3	132.4	248.2	413.8	744.8
0	19.0	49.7	99.3	132.4	248.3	413.8	744.8
38	19.0	49.7	99.3	132.4	248.3	413.8	744.8
50	18.4	48.1	96.3	128.4	240.6	401.0	722.9
100	16.0	42.3	84.5	112.8	211.0	351.7	633.2
150	14.8	38.6	77.1	102.9	192.4	320.9	577.7
200	13.6	35.8	71.2	95.2	178.5	297.3	535.2
250	12.0	33.5	66.8	89.1	167.1	278.3	501.0
300	10.2	31.6	63.1	84.5	158.1	263.6	474.5
350	8.4	30.4	61.0	81.3	152.3	253.8	456.9
375	7.4	29.6	59.9	79.8	149.3	249.1	448.3
400	6.5	29.3	58.9	78.6	147.2	245.4	441.9
425	5.6	29.0	58.3	77.6	145.6	242.9	437.2
450	4.6	29.0	57.7	77.0	144.4	240.4	432.8
475	3.7	28.7	57.3	76.4	143.4	239.0	430.3
500	2.8	27.3	54.8	72.9	136.7	228.0	410.5
525	1.9	25.2	50.7	67.5	126.4	210.7	379.2
550	1.4	24.0	47.8	63.9	119.8	199.5	359.0
575	1.4	22.9	45.5	60.8	114.0	190.1	341.9
600	1.4	19.9	39.8	53.1	99.5	166.0	298.6
625	1.4	15.7	31.7	42.1	79.2	131.8	237.3
650	1.4	12.6	25.3	33.8	63.3	105.7	189.8
675	1.4	10.2	20.7	27.3	51.5	86.1	154.8
700	1.4	8.3	16.9	22.3	42.0	69.8	125.8
725	1.4	6.9	13.9	18.6	35.0	58.2	104.9
750	1.4	5.7	11.3	15.2	28.7	47.7	85.6
775	1.4	4.6	9.0	12.1	22.9	38.0	68.4
800	1.4	3.5	7.0	9.3	17.4	29.2	52.6
816	1.4	2.8	5.9	7.6	14.1	23.8	42.8

Notes: 1 MPa = 10 bar; 100 kPa = 1 bar

Figure B–7 *Standard components of construction for forged small-bore gate, globe, and check valves. (Printed with the kind permission of OMB Valves, spa, Italy.)*

Appendix B—General Engineering Data   227

1 inch = 25,4 mm

Inch	0	1	2	3	4	5	6	7	8	9	10	11	12	13	14
0		0.000	25.400	50.800	76.200	101.60	127.00	152.40	177.80	203.20	228.60	254.00	279.40	304.80	330.20	355.60
1/64		0.397	25.797	51.197	76.597	102.00	127.40	152.80	178.20	203.60	229.00	254.40	279.80	305.20	330.60	356.00
1/32		0.794	26.194	51.594	76.994	102.39	127.79	153.19	178.59	203.99	229.39	254.79	280.19	305.59	330.99	356.39
3/64		1.191	26.591	51.991	77.391	102.79	128.19	153.59	178.99	204.39	229.79	255.19	280.59	305.99	331.39	356.79
1/16		1.587	26.987	52.387	77.787	103.19	128.59	153.99	179.39	204.79	230.19	255.59	280.99	306.39	331.79	357.19
5/64		1.984	27.384	52.784	78.184	103.58	128.98	154.38	179.78	205.18	230.58	255.98	281.38	306.78	332.18	357.58
3/32		2.381	27.781	53.181	78.581	103.98	129.38	154.78	180.18	205.58	230.98	256.38	281.78	307.18	332.58	357.98
7/64		2.778	28.178	53.578	78.978	104.38	129.78	155.18	180.58	205.98	231.38	256.78	282.18	307.58	332.98	358.38
1/8		3.175	28.575	53.975	79.375	104.77	130.17	155.57	180.97	206.37	231.77	257.17	282.57	309.97	333.37	358.77
9/64		3.572	28.972	54.372	79.772	105.17	130.57	155.97	181.37	206.77	232.17	257.57	282.97	308.37	333.77	359.17
5/32		3.969	29.369	54.769	80.169	105.57	130.97	156.37	181.77	207.17	232.57	257.97	283.37	308.77	334.17	359.57
11/64		4.366	29.766	55.166	80.566	105.97	131.37	156.77	182.17	207.57	232.97	258.37	283.77	309.17	334.57	359.97
3/16		4.762	30.162	55.562	80.962	106.38	131.76	157.16	182.56	207.96	233.36	258.76	285.16	309.56	334.96	360.38
13/64		5.159	30.559	55.959	81.359	106.76	132.16	157.56	182.96	208.36	233.76	259.16	284.56	309.96	335.36	360.76
7/32		5.556	30.956	56.356	81.756	107.16	132.56	157.96	183.36	208.76	234.16	259.56	284.96	310.36	335.76	361.16
15/64		5.953	31.353	56.753	82.153	107.55	132.95	158.35	183.75	209.15	234.55	259.95	285.35	310.75	336.15	361.55
1/4		6.350	31.750	57.150	82.550	107.95	133.35	158.75	184.15	209.55	234.95	260.35	285.75	311.15	336.55	361.95
17/64		6.747	32.147	57.547	82.947	108.35	133.75	159.15	184.55	209.95	235.35	260.75	286.15	311.55	336.95	362.35
9/32		7.144	32.544	57.944	83.344	108.74	134.14	159.54	184.94	210.34	235.74	261.14	286.54	311.94	337.34	362.74
19/64		7.541	32.941	58.341	83.741	109.14	134.54	159.94	185.34	210.74	236.14	261.54	286.94	312.34	337.74	363.14
5/16		7.937	33.337	58.737	84.137	109.54	134.94	160.34	185.74	211.14	236.54	261.94	287.34	312.74	338.14	363.54
21/64		8.334	33.734	59.134	84.534	109.93	135.33	160.73	186.13	211.53	236.93	262.33	287.73	313.13	338.53	363.93
11/32		8.731	34.131	59.531	84.931	110.33	135.73	161.13	186.53	211.93	237.33	262.73	288.13	313.53	338.93	364.33
23/64		9.128	34.528	59.928	85.328	110.73	136.13	161.53	186.93	212.33	237.73	263.13	288.53	313.93	339.33	364.73
3/8		9.525	34.925	60.325	85.725	111.12	136.52	161.92	187.32	212.72	238.12	263.52	288.92	314.32	339.72	365.12
25/64		9.922	35.322	60.722	86.122	111.52	136.92	162.32	187.72	213.12	238.52	263.92	289.32	314.72	340.12	365.52
13/32		10.319	35.719	61.119	86.519	111.92	137.32	162.72	188.12	213.52	238.92	264.32	289.72	315.12	340.52	365.92
27/64		10.716	36.116	61.516	86.916	112.32	137.72	163.12	188.52	213.92	239.32	264.72	290.12	315.52	340.92	366.32
7/16		11.112	36.512	61.912	87.312	112.71	138.11	163.51	188.91	214.31	239.71	265.11	290.51	315.91	341.31	366.71
29/64		11.509	36.909	62.309	87.709	113.11	138.51	163.91	189.31	214.71	240.11	265.51	290.91	316.31	341.71	367.11
15/32		11.906	37.306	62.706	88.106	113.51	138.91	164.31	189.71	215.11	250.41	265.91	291.31	316.71	342.11	367.51
31/64		12.303	37.703	63.103	88.503	113.90	139.30	164.70	190.10	215.50	240.90	266.30	291.70	317.10	342.50	367.90
1/2		12.700	38.100	63.500	88.900	114.30	139.70	165.10	190.50	215.90	241.30	266.70	292.10	317.50	342.90	368.30
33/64		13.097	38.497	63.897	89.297	114.70	140.10	165.50	190.90	216.30	241.70	267.10	292.50	317.90	343.30	368.70
17/32		13.494	38.894	64.294	89.694	115.09	140.49	165.89	191.29	216.69	242.09	267.47	292.89	318.29	343.69	369.09
35/64		13.891	39.291	64.691	90.091	115.49	140.89	166.29	191.69	217.09	242.49	267.89	293.29	318.69	344.09	369.49
9/16		14.287	39.687	65.087	90.487	115.89	141.29	166.69	192.09	217.49	242.89	268.29	293.69	319.09	344.49	369.89
37/64		14.684	40.084	65.484	90.884	116.28	141.68	167.08	192.48	217.88	243.28	268.68	294.08	319.48	344.88	370.28
19/32		15.081	40.481	65.881	91.281	116.68	142.08	167.48	192.88	218.28	243.68	269.08	294.48	319.88	345.28	370.68
39/64		15.478	40.878	66.278	91.678	117.08	142.48	167.88	193.28	218.68	244.08	269.48	294.88	320.28	345.68	371.08
5/8		15.875	41.275	66.675	92.075	117.47	142.87	168.27	193.67	219.07	244.47	269.87	295.27	320.67	346.07	371.47
41/64		16.272	41.672	67.072	92.472	117.87	143.27	168.67	194.07	219.47	244.87	270.27	295.67	321.07	346.47	371.87
21/32		16.669	42.069	67.469	92.869	118.27	143.67	169.07	194.47	219.87	245.27	270.67	296.07	321.47	346.87	372.27
43/64		17.066	42.466	67.866	93.266	118.67	144.07	169.47	194.87	220.27	245.67	271.07	296.47	321.87	347.27	372.67
11/16		17.462	42.862	68.262	93.662	119.06	144.46	169.86	195.26	220.66	246.06	271.46	296.86	322.26	347.66	373.06
45/64		17.859	43.259	68.659	94.059	119.46	144.86	170.26	195.66	221.06	246.46	271.86	297.26	322.66	348.06	373.46
23/32		18.256	43.656	69.056	94.456	119.86	145.26	170.66	196.06	221.46	246.86	272.26	297.66	323.06	348.46	373.86
47/64		18.635	44.053	69.453	94.853	120.25	145.65	171.05	196.45	221.85	247.25	272.65	298.05	323.45	348.85	374.25
3/4		19.050	44.450	69.850	95.250	120.65	146.05	171.45	196.85	222.25	247.65	273.05	298.45	323.85	349.25	374.65
49/64		19.447	44.487	70.247	95.647	121.05	146.45	171.85	197.25	222.65	248.05	273.45	298.85	324.25	349.65	375.05
25/32		19.844	45.244	70.644	96.044	121.44	146.84	172.24	197.64	223.04	248.44	273.84	299.24	324.64	350.04	375.44
51/64		20.241	45.641	71.041	96.441	121.84	147.24	172.64	198.04	223.44	248.84	274.24	299.64	325.04	350.44	375.84
13/16		20.637	46.037	71.437	96.837	122.24	147.64	173.04	198.44	223.84	249.24	274.64	300.04	325.44	350.84	376.24
53/64		21.034	46.434	71.834	97.234	122.63	148.03	173.43	198.83	224.23	249.63	275.03	300.43	325.83	351.23	376.63
27/32		21.431	46.831	72.231	97.631	123.03	148.43	173.83	199.23	224.63	250.03	275.43	300.83	326.23	351.63	377.03
55/64		21.828	47.228	72.628	98.028	123.43	148.83	174.23	199.63	225.03	250.43	275.83	301.23	326.63	352.03	377.43
7/8		22.225	47.625	73.025	98.425	123.82	149.22	174.62	200.02	225.42	250.82	276.22	301.62	327.02	352.42	377.82
57/64		22.622	48.022	73.422	98.822	124.22	149.62	175.02	200.42	225.82	251.22	276.62	302.02	327.42	352.82	378.22
29/32		23.019	48.419	73.819	99.219	124.62	150.02	175.42	200.82	226.22	251.62	277.02	302.42	327.82	353.22	378.62
59/64		23.416	48.816	74.216	99.616	125.02	150.42	175.82	201.22	226.62	252.02	277.42	302.82	328.22	353.62	379.02
15/16		23.812	49.212	74.612	100.01	125.41	150.81	176.21	201.61	227.01	252.41	277.81	303.21	328.61	354.01	379.41
61/64		24.209	49.609	75.009	100.41	125.81	151.21	176.61	202.01	227.41	252.81	278.21	303.61	329.01	354.41	379.81
31/32		24.606	50.006	75.406	100.81	126.21	151.61	177.01	202.41	227.81	253.21	278.61	304.01	329.41	354.81	380.21
63/64		25.003	50.403	75.803	101.20	126.60	152.00	177.40	202.80	228.20	253.60	279.00	304.40	329.80	355.20	380.63

Figure B–8 Conversion chart dimensions—U.S. customary units to metric units. (Printed with kind permission of OMB Valves, spa, Italy.)

228 Appendix B—General Engineering Data

GATE VALVE

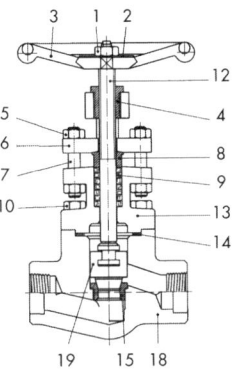

1 Wheelnut
2 Nameplate
3 Handwheel
4 Yoke Nut
5 Gland Nut
6 Gland Flange
7 Gland Stud
8 Gland
9 Packing
10 Bolts
12 Stem
13 Bonnet
14 Gasket
15 Seat
17 Wedge
18 Body

GLOBE VALVE

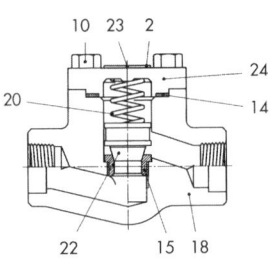

1 Wheelnut
2 Nameplate
3 Handwheel
4 Yoke Nut
5 Gland Nut
6 Gland Flange
7 Gland Stud
8 Gland
9 Packing
10 Bolts
12 Stem
13 Bonnet
14 Gasket
15 Seat
18 Body
19 Disc

CHECK VALVE

2 Nameplate
10 Bolts
14 Gasket
15 Seat
18 Body
20 Spring
22 Piston
23 Rivet
24 Cap

Figure B–9 *Standard components of construction for forged small-bore gate, globe, and check valves. (Printed with the kind permission of OMB Valves, spa, Italy.)*

GATE VALVE

	A105/F6	A105/F6HFS	LF2/304	F11/F6HFS	F304/304	F316/316
Wheelnut	Carbon Steel	Carbon Steel	Carbon Steel	Carbon Steel	Carbon Steel	Carbon Steel
Nameplate	Aluminium	Aluminium	Aluminium	Aluminium	Aluminium	Aluminium
Handwheel	Carbon Steel	Carbon Steel	Carbon Steel	Carbon Steel	Carbon Steel	Carbon Steel
Yoke Nut	416	416	416	416	303	303
Gland Nut	2H	2H	GR8	GR8	GR8	GR8
Gland Flange	A105	A105	F6	F6	F304	F304
Gland Stud	410	410	B8	B8	B8	B8
Gland	316L	316L	316L	316L	316L	316L
Packing (*)	Graphite	Graphite	Graphite	Graphite	Graphite	Graphite
Bolts	B7	B7	L7	B16	B8	B8
Stem	410	410	304	410	304	316
Bonnet	A105	A105	LF2	F11	F304	F316
Gasket	Sp. Wound	Sp. Wound	Sp. Wound	Sp. Wound	Sp. Wound	Sp. Wound
Seat	410	410HF	304	410HF	304	316
Wedge	F6	F6	F304	F6	F304	F316
Body	A105	A105	LF2	F11	F304	F316

GLOBE VALVE

	A105/F6	A105/F6HFS	LF2/304	F11/F6HFS	F304/304	F316/316
Wheelnut	Carbon Steel	Carbon Steel	Carbon Steel	Carbon Steel	Carbon Steel	Carbon Steel
Nameplate	Aluminium	Aluminium	Aluminium	Aluminium	Aluminium	Aluminium
Handwheel	Carbon Steel	Carbon Steel	Carbon Steel	Carbon Steel	Carbon Steel	Carbon Steel
Yoke Nut	416	416	416	416	303	303
Gland Nut	2H	2H	GR8	GR8	GR8	GR8
Gland Flange	A105	A105	F6	F6	F304	F304
Gland Stud	410	410	B8	B8	B8	B8
Gland	316L	316L	316L	316L	316L	316L
Packing (*)	Graphite	Graphite	Graphite	Graphite	Graphite	Graphite
Bolts	B7	B7	L7	B16	B8	B8
Stem	410	410	304	410	304	316
Bonnet	A105	A105	LF2	F11	F304	F316
Gasket	Sp. Wound	Sp. Wound	Sp. Wound	Sp. Wound	Sp. Wound	Sp. Wound
Seat	410	410HF	304	410HF	304	316
Disc	410	410	304	410	304	316
Body	A105	A105	LF2	F11	F304	F316

CHECK VALVE

	A105/F6	A105/F6HFS	LF2/304	F11/F6HFS	F304/304	F316/316
Nameplate	Aluminium	Aluminium	Aluminium	Aluminium	Aluminium	Aluminium
Bolts	B7	B7	L7	B16	B8	B8
Gasket	Sp. Wound	Sp. Wound	Sp. Wound	Sp. Wound	Sp. Wound	Sp. Wound
Seat	410	410HF	304	410HF	304	316
Body	A105	A105	LF2	F11	F304	F316
Spring	Arm. Steel	Arm. Steel	Arm. Steel	Arm. Steel	Arm. Steel	Arm. Steel
Piston	410	410	304	410	304	316
Rivet	Carbon Steel	Carbon Steel	Carbon Steel	Carbon Steel	Carbon Steel	Carbon Steel
Cap	A105	A105	LF2	F11	F304	F316

Figure B–10   *Standard materials of construction for small-bore gate, globe, and check valves. (Printed with the kind permission of OMB Valves, spa, Italy.)*

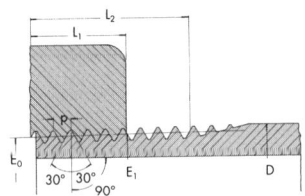
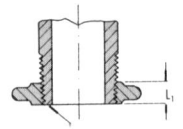
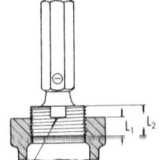

Flush by Hand
Tolerance on Product
One turn large or small from notch on plug gauge or face of ring gauge.
Notch flush with face of fitting. If chamfered, notch flush with bottom of chamfer.

$E_0 = D - (0.050D + 1.1)p$ p = Pitch
*$E_1 = E_0 + 0.0625 L_1$ Depth of thread = 0.80p
$L_2 = (0.80D = 6.8)p$ Total Taper ¾-inch per Foot

THREADS - ASME B1.20.1

Nominal pipe size	D Outside diameter of pipe	Number of threads per inch	p Pitch of thread	E_0 Pitch diameter at end of external thread	E_1 ■ Pitch diameter at end of external	L_1 ● Normal engagement by hand between external and internal threads	L_2 ♦ Length of effective external thread	Height of thread
¹⁄₁₆	0.3125	27	0.03704	0.27118	0.28118	0.160	0.2611	0.02963
⅛	0.405	27	0.03704	0.36351	0.37360	0.1615	0.2639	0.02963
¼	0.540	18	0.05556	0.47739	0.49163	0.2278	0.4018	0.04444
⅜	0.675	18	0.05556	0.61201	0.62701	0.240	0.4078	0.04444
½	0.840	14	0.07143	0.75843	0.77843	0.320	0.5337	0.05714
¾	1.050	14	0.07143	0.96768	0.98887	0.339	0.5457	0.05714
1	1.315	11.5	0.08696	1.23863	1.23863	0.400	0.6828	0.06957
1¼	1.660	11.5	0.08696	1.55713	1.58338	0.420	0.7068	0.06957
1½	1.900	11.5	0.08696	1.79609	1.82234	0.420	0.7235	0.06957
2	2.375	11.5	0.08696	2.26902	2.29627	0.436	0.7565	0.06957

■ Also pitch diameter at gauging notch.
♦ Also length of plug gauge.
● Also length of ring gauge, and length from gauging notch to small end of plug gauge.
* For the ¹⁄₁₆-27 and ¹⁄₄-18 sizes... E_1 approx. = $D - (0.05D + 0.827)$ p.

DEPTH

Figure B–11   Threaded end connections to ASME B1.20.1. (Printed with the kind permission of OMB Valves, spa, Italy.)

Index

Numerics
100% examination 196
100% radiography 199

A
alignment 186
alloying 117
aluminum (Al) 122
American Iron and Steel Institute (AISI) 25
American National Standards Institute (ANSI) 14
American Petroleum Institute (API) 25, 30
American Petroleum Institute's API 5L 63
American Society for Nondestructive Testing (ASNT) 25, 39
American Society for Quality 39
American Society for Testing and Materials 25, 32–39, 133
American Society of Mechanical Engineers (ASME) 4, 25
American Water Works Association (AWWA) 25, 41
American Welding Society (AWS) 25, 40

annealing 131
   full annealing 131
   process annealing 131
   spheroidizing annealing 131
ASME
   B1.20.1, 1983, Pipe Threads, General Purpose, Inch 67
   B16.25, Butt Welding Ends 70
   B16.34, Valves—Flanged, Threaded, and Welding End 94
   B16.47, Large Diameter Steel Flanges 80
   B16.5, 2003, Pipe Flanges and Flanged Fittings 79
   B31, Codes for Pressure Piping 191
   B31.3, 116, 186
   B31.3, Process Piping Table 331.1 182
   B31.4, Pipeline Transportation Systems for Liquid 175
   B31.8, Gas Transportation and Distribution Piping Systems 175

ASME, *continued*
B36.10M, Welded and
   Seamless Wrought Steel
   Pipe 63
B36.19M, Stainless Steel
   Pipe 63
ASME Boiler and Pressure Vessel
   Code 4, 191
Boilers and Pressure Vessels 13
Fiber-Reinforced Plastic
   Pressure Vessels 12
Material Specifications 5, 6
Nondestructive
   Examination 10
Nuclear Components 13
Power Boilers 5
Recommended Guidelines
   for the Care of Power
   Boilers 10
Recommended Rules for the
   Care and Operation of
   Heating Boilers 10
Rules for Construction and
   Continued Service of
   Transport Tanks 13
Rules for Construction of
   Heating Boilers 9
Rules for Construction of
   Nuclear Facility
   Components 7–9
Rules for Construction of
   Pressure Vessels 10, 11
Rules for In-Service
   Inspection of Nuclear
   Power Plant
   Components 12
Welding and Brazing
   Qualifications 12
ASME Codes and Standards
   flanges 103
ASME Codes for Pressure Piping 14
   Building Services Piping 18
   Fuel Gas Piping 15
   Gas Transmission and
      Distribution Piping
      Systems 17
   Managing System Integrity
      of Gas Pipelines 18
   Manual for Determining
      Remaining Strength of
      Corroded Pipelines 19
   Pipeline Transportation
      Systems for Liquid
      Hydrocarbons and Other
      Liquids 16
   Power Piping 15
   Process Piping 15, 19
   Refrigeration Piping and
      Heat Transfer
      Components 16
   Slurry Transportation Piping
      Systems 18
ASME Guide for Gas
   Transmission and
   Distribution Piping
   Systems 19
ASME Standard 26
   Face-to-Face and End-to-End
      Dimensions of Valves 28
   Factory-Made Wrought Butt
      Welding Fittings 28
   Forged Fittings, Socket
      Welding and Threaded 28
   Large Diameter Steel
      Flanges 29
   Malleable Iron Threaded Pipe
      Unions 29
   Metallic Gaskets for Pipe
      Flanges, Ring Joint Spiral
      Wound and Jacketed 28
   Nonmetallic Flat Gaskets for
      Pipe Flanges 29
   Orifice Flanges 29
   Pipe Flanges and Flanged
      Fittings
      NPS ½ through 24 27

Pipe Threads, General
Purpose, Inch 27
Stainless Steel Pipe 30
Steel Line Blanks 30
Valves Flanged, Threaded
and Welding End 29
Welded and Seamless
Wrought Steel Pipe 30
assembly and erection 177
autogenous welds 180
AWWA Standards
Concrete Pipe 42
Ductile-Iron Pipe and
Fittings 41
Plastic Pipe 42
Steel Pipe 42
Valves and Hydrants 42

B
backseat 101
bending 175
buckling 175
and forming 177
ovality 175
thinning 175
beveled ends (BE) 72
beveling 174
boiler pressure vessel (BPV) 3
bolt 110
bolting 109
and gaskets 101
and nut selection 110
tightening 112
boron (B) 122
brazing and soldering 177, 183
Brownfield project 158
buckling 175
butt-weld end fittings 72

C
camprofile 108
Canadian Standards Association 43

carbon (C) 122
chemical properties of metals 117
chromium (Cr) 123
circumferential butt welds 196
civil engineering 155
closing element 98
code 2
    ASME B31.3 3
codes and standards
    considerations 172
columbium (Nb) 123
Combination Hydrostatic-
    Pneumatic Leak Test 204
commissioning 169
commissioning phase 169
compressed asbestos fiber (CAF)
    gaskets
    types of gaskets 107
compressed gas association 43
computer-aided design (CAD) 1
conception phase 160
construction 13, 166
construction phase 166
continued service 13
copper (Cu) 123
Copper Development
    Association 42
Copper Tube Handbook 42
corrosion allowance (CA) 66, 184
corrosion-resistant alloys (CRA) 62
cutting 174

D
density 121
design code
    ASME B31.3 24
detailed engineering 162
detailed engineering phase 162
diamètre nominal (DN) 52, 63, 64,
    92
double random length (DRL) 61
dye penetrant inspection (DPI) 75

E

engineering and procurement (E&P) 162
engineering, procurement, and construction (EPC) 137, 162
erection 171
exotic materials 66
Expansion Joint Manufacturers Association 43

F

fabrication 171
fasteners 113
feasibility phase 160
ferrous materials 122
flange 74, 103
    blind 78
    dimensional standards 79
    lap-joint 78, 188
    screwed 188
    slip-on 78, 188
    socket-weld 76, 188
    threaded 77
    weld-neck 188
    weldneck 75
flange faces 104
    flat face 105
    raised face 104
    ring-type oint 104
flange specification 105
    flange pressure class 105
    material 106
    nominal pipe size 105
    standard 106
    type and facing 105
flanged joint system 74, 111, 188
    assembly 113
    bolt condition 112
    flange condition 111
    gasket condition 111
    gasket quality 112
flux 180
forming 175

front-end engineering development (FEED) phase 161

G

gardening 132
gasket 106
glass-reinforced epoxy (GRE) 51
glass-reinforced plastic (GRP) 51
grade 21
grain 121
    boundaries 121
    size 121
grassroots project 157

H

hardenability 121
hardness 119
health, safety, and environmental (HSE) requirements 1
heat treatment 176, 181
heat-affected zone (HAZ) 181
high-carbon steel 125
high-tensile steel 125
hot cuts 174
hydrostatic testing 200, 202

I

industrial piping systems testing limitations 201
industrial pressure piping
    design 194
    fabrication 194, 196
    inspection 194
    testing 200
initial service testing 200, 205
inside diameter (ID) 52
installation 171
instrument engineering 155
iron (Fe) 123
iron pipe size (IPS) 62
issued for construction (IFC) 162

J
joint integrity 102

K
Kammprofile gaskets 108

L
lead (Pb) 123
leakage field 198
liquid penetrant examination 197
longitudinal welds 53, 196
low-temperature carbon steel (LTCS) 66

M
magnetic particle examination (MPE) 198
magnetic particle inspection (MPI) 75
manganese (Mn) 123
Manufacturers Standardization Society (MSS) 25
   of the Valve and Fittings Industry 25, 43
mechanical engineering 154
mechanical properties of metals 118
modulus of elasticity 119
molybdenum (Mo) 123

N
National Association of Corrosion Engineers (NACE) 25, 45
National Fire Protection Association (NFPA) 25
National Fire Codes 45
National Fire Protection Association (NFPA) 45
necks down 119
nickel (Ni) 123
nominal pipe size (NPS) 52, 64
nondestructive examination (NDE) 72
nonferrous materials 122, 132
normalizing 132
nuts 110

O
outside diameter (O.D.) 52
ovality 175

P
phase 161, 162, 166
   construction 166
   detailed engineering 162
   front-end engineering development 161
   precommissioning and commissioning 169
phonographic 76
phosphorus (Ph) 124
physical properties of metals 120
pipe 63
   butt-weld ends 70
   cladding 186
   dimensional specifications 63
   fabrication 172
   fabrication 174
   fittings 70
   galvanizing 184
   material specifications 66
   sizes 62
   threaded ends 67
pipe ends 67
   dimensional standards 67
Pipe Fabrication Institute (PFI) 25, 46
pipeline system 18
piping
   design group 142
   flexibility analysis 168
   isometrics 168
   materials control group 145

piping, *continued*
  stress engineering group 148
  system 50, 51
pneumatic testing 200, 204
precommissioning 169
precommissioning and
  commissioning phase 169
precommissioning phase 169
preheating 176, 181
process
  engineering 151, 153
process flow diagrams (PFDs) 152
project lead piping
  engineer (PEL) 137, 138
  materials controller 146
  materials engineer 140
  stress engineer 148
project phases 160
project piping
  area/unit supervisor (squad boss) 143
  CAD coordinator 143
  designers-checkers 144
  materials controller 147
  stress engineer 148
project types 157
punch list 206

R
radiographic examination 199
random examination 196
random radiography 200
random spot examination 197
reprepped 67
requests for quotation (RFQs) 165
required-on-site (ROS) date 166
revamp 158

S
schedule (SCH) 62
senior piping materials engineer 141
silicon (Si) 124

single random length (SRL) 61
Society of Automotive Engineers (SAE) 25, 48
specific heat 121
specification 2
spiral welding 60
spot examination 196
spot radiography 200
squad bosses 142
square cut 67
stainless steel 126
  austenitic 126, 127
  duplex grades 128
  ferritic grades 128
  martensitic 127
  precipitation hardenable 129, 130
  super stainless steels 130
standard 2
  ASME B16.5 2
standard specifications 2
  ASTM A105 2
startup and handover to the owner 169
steel
  medium-carbon 125
  mild (low-carbon) 125
stem packing 100
stem protector 101
stress isometrics 168
stress relief 131
structural engineering 155
sulfur (S) 124

T
technical bid evaluation (TBE) 165
tempering 132
test pack 206
thermal conductivity 121
thermal expansion 121
thinning 175
threaded joints 188
titanium (Ti) 124

Index 237

toughness 120
tube 63
tungsten (W) 124
types of gaskets 106
   compressed nonasbestos
      fiber (CNAF) 107
   nonmetallic gasket 107
   semi-metallic gaskets 107

U
ultimate tensile strength (UTS) 118
ultrasonic examination 200
unified numbering system
      of ferrous metals and
      alloys 135
unlisted material 66

V
vacuum testing 200
valve 80, 81
   API specification 83
   API standards 84
   automatic 87
   AWWA standards and
      specifications 83
   codes and standards 82
   linear 87
   MSS standards 84
   non-pressure-containing
      parts 99–101
   plug-type 81
   pressure class ratings 92
   pressure containing parts 95
   quarter-turn 87
   rotary 87
   seat 99
   size 92
   small-bore 92
   valve stem 99
      nonrising stem with inside
         screw 100
      rising stem with inside
         screw 100
      rising stem with outside
         screw and yoke 99
      rotary stem 100
      sliding stem 100
vanadium (V) 124
visual examination 197

W
welding 176, 178
   gas metal arc (GMAW) 180
   gas shielded arc 180
   gas tungsten arc 180
   metal inert gas (MIG) 180
   shielded metal arc 179
   submerged arc 180
   tungsten inert gas (TIG) 180
   weld repair 181

Y
yield strength 119
yoke 101
Young's modulus 119